T5-DGQ-320

The Soviet Union and its Geographical Problems

Also by Roy E. H. Mellor

Geography of the USSR
Sowjetunion
Comecon – Challenge to the West
Eastern Europe : A Geography of the Comecon Countries
The Two Germanies : A Modern Geography
Europe: A Geographical Survey of the Continent (with E. A. Smith)

The Soviet Union and its Geographical Problems

Roy E. H. Mellor
Department of Geography, University of Aberdeen

First published 1982 by
THE MACMILLAN PRESS LTD
London and Basingstoke
Companies and representatives throughout the world

ISBN 0 333 27662 0 (hard cover)
ISBN 0 333 27663 9 (paper cover)

Typeset by
CAMBRIAN TYPESETTERS
Farnborough, Hants

Printed in Hong Kong

To H. K. M.

Contents

Introduction xi
List of Maps xiii
List of Tables xiv

1 The Soviet Milieu 1

1.1 The major landscape zones 8
The tundra 8
The tayga 10
The mixed forests 13
The steppe 14
The desert 15
The Transcaucasian lowlands 16
The mountains 16
1.2 Environment and the Soviet milieu 17
Conservation and pollution problems in the Soviet setting 21
1.3 Where to follow up this chapter 23

2 The Soviet State – its Territorial Origins and its Internal Organisation 24

2.1 Territorial growth – the beginnings 25
2.2 Territorial growth from the eighteenth century 28
2.3 Territorial problems after the Revolution of 1917 29
2.4 Territorial gains after the Second World War 30
2.5 Internal territorial problems 32
2.6 The territorial-administrative organisation 34
2.7 Economic regionalisation 38
2.8 Where to follow up this chapter 45

3 **The Soviet People** 46

3.1 The ethnic problem 50
3.2 Geographical distribution of population 57
3.3 The role of migration in the population pattern 66
3.4 Where to follow up this chapter 70

4 **Settlements – Where Soviet People Live** 71

4.1 The village 71
4.2 The town 73
4.3 Where to follow up this chapter 83

5 **Soviet Agriculture** 84

5.1 Where to follow up this chapter 101

6 **Soviet Industry – The Resource Base** 102

6.1 Energy resources 105
6.2 Spacial distribution of the main energy resources 107
Coal 107
Oil 110
Electricity 113
6.3 Metallic minerals 116
6.4 Iron and the ferrous metals 118
6.5 Non-ferrous metals 119
6.6 Minerals for the chemicals industry 120
6.7 Where to follow up this chapter 121

7 **Soviet Industry – The Major Branches** 122

7.1 The metallurgical industries 128
7.2 Non-ferrous metallurgy 132
7.3 Engineering 134
7.4 The chemicals industry 138
7.5 Textile industries 141
7.6 Where to follow up this chapter 144

8 **Transport – Holding the Soviet Economy Together** 146

8.1 Railways 152
8.2 Inland waterways 159
8.3 Sea-going shipping 162
8.4 Road transport 165
8.5 Pipelines 167
8.6 Air transport 167

8.7 Postal- and telecommunications 168
8.8 Where to follow up this chapter 169

9 The Soviet Union and the World since 1945 170

9.1 The Soviet trading pattern 173
9.2 Changing spatial patterns in the USSR 180
9.3 The Socialist *bloc* in the world 183
9.4 The Soviet dilemma in the Third World 184
9.5 The Soviet strategic position 187
9.6 Towards the year 2000 192
9.7 Where to follow up this chapter 194

Bibliography 197
Index 201

Introduction

This review of the geography of the Soviet Union is designed for students and others seeking an introduction to this, the world's largest compact political territory, encompassing one-sixth of the land surface of our planet. In assessing the achievements or failures of the USSR or in making comparisons between it and other powers, it is essential to appreciate the real nature of the environment, both in terms of its physical setting and its time–space matrix, in which Soviet citizens live out their lives. In so many respects the Soviet *milieu* is bedevilled by the harshness of the physical world around it and by the sheer 'embarrassment of space'.

The study is built around nine chapters, each devoted to a specific thematic topic in the geography of the country as a whole. In the contemporary setting it is felt important to concentrate attention on an overview of the territory as a whole rather than follow the conventional and perhaps now rather dated plan of a brief general review and then a region-by-region survey. For the majority of people seeking a general overview of the Soviet Union, the coherence of the whole is most meaningful.

The Soviet Union lies upon a threshold of rapid change. The pre-war and early post-war economy was built along lines that resembled nineteenth-century industrialisation in Western Europe. Now, having increasingly entered the arena of world trade and technology, a new sophistication is being added by the turbulent effort to keep abreast of the accelerating technological change brought about by the age of the silicon chip that has dawned in the Western world. At the same time, patterns of demographic change pose challenges in the supply of labour and in internal political forces through the shift in balance between the major ethnic groups. As a new 'window on the West' has opened, consumer pressures for a bigger slice of the national cake have also developed.

There has also been the relaxation of the tight encapsulation of Stalinist days, not only in the attitude to the world at large, but also as the Council for Mutual Economic Assistance (*Comecon*) has evolved. It is now unreal to consider the geography of the Soviet Union without reference to the country's role as the core of this group of socialist nations acting in concert for mutual economic advantage. This is examined in particular in the final chapter, where an assessment of the world position of the Soviet Union is attempted. Such a review represents one 'model' of the world scene; readers may care to construct others for themselves.

The text owes much to the useful comments made by my colleague, E. Alistair Smith. The manuscript was typed by Jane Calder, the maps prepared by Laurie McLean and Philip Glennie, while Sheila Bain helped in the collection of material. My wife read the proofs, prepared the index and kept the 'inner man' in good fettle while writing. To everybody I am most grateful. I must also record the pleasure of working with Derick Mirfin, John Winckler, Keith Povey and Steven Kennedy, 'Macmillan's men'.

Old Aberdeen
June 1981

Roy E. H. Mellor

List of Maps

1.1 The world position of the Soviet Union 2
1.2 Terrain types 5
1.3 Major natural landscapes 9
2.1 The growth of Russia 27
2.2 The western boundaries of the Soviet Union 31
2.3 Administrative-territorial structure 35
2.4 Boundaries of planning regions 40
3.1 Age and sex structure of the Soviet Union, 1970 48
3.2 The ethnic map of the USSR 52
3.3 Population distribution in the USSR 58
4.1 The distribution of Soviet urban population 75
5.1 The physical conditions of farming 86
5.2 Types of farming in the USSR 94
6.1 Mineral deposits in the USSR 103
6.2 Energy resources in the USSR 106
6.3 Electricity supply in the USSR 114
7.1 The distribution of industry in the USSR 142
8.1 The Soviet concept of a 'unified transport system' 147
8.2 The Soviet railway system 153
9.1 The world distribution of Soviet aid 179

List of Tables

3.1 Growth of population 47
3.2 Births, deaths and natural increase by republics per 1,000 of the population 49
3.3 Ethnic structure of the USSR 51
4.1 Settlements of town type – criteria for definition in selected republics 78
6.1 Soviet mineral production and its world share 104
7.1 Position of the USSR in world industrial production 144
8.1 Division of transport effort between the media 148
8.2 Division of originating goods and passengers among the transport media 151
9.1 Soviet foreign trade (in million roubles at contemporary prices) 175

1

The Soviet Milieu

Any traveller in the Soviet Union is quickly impressed by the immensity and the harshness of the physical environment that forms the backcloth to the social and economic milieu. In thinking of Soviet problems, Europeans must come to terms with dimensions far greater than they experience in their own homelands, though North Americans may find the contrast less striking. Yet if our perception of the behavioural environment of the Soviet people is to be reasonably accurate, we must try to appreciate the real immensity and hostility of the physical environment in which they act out their everyday lives.

With an area of 22,402,200km^2 (8,647,250 sq. miles), the Soviet Union is the world's largest country, occupying one-sixth of the land surface of our planet (Figure 1.1). It is perhaps easier to appreciate this by saying it is ninety times the area of the United Kingdom and over twice that of the USA. It extends across almost half the circumference of the earth, from roughly 20°E near Kaliningrad on the Baltic to just beyond 170°W at Cape Dezhnev in Pacific waters. This great east–west extent creates a massive internal problem of communication by straddling eleven time zones: when people in Moscow are going to bed, those in Vladivostok are just getting up! The maximum north–south distance of some 4,800km (3,000 miles) looks more modest, but still gives rise to the environmental implications of extending from the arctic wastes north of 77°N near Cape Chelyushkin across the more temperate latitudes deep into the arid interior of Asia to the foot of the Central Asian mountains, roughly about 35°N. The Soviet Union also claims jurisdiction over a vast sector of the Arctic Ocean basin between just west of Murmansk to the Bering Strait, with the apex lying at the North Pole itself.

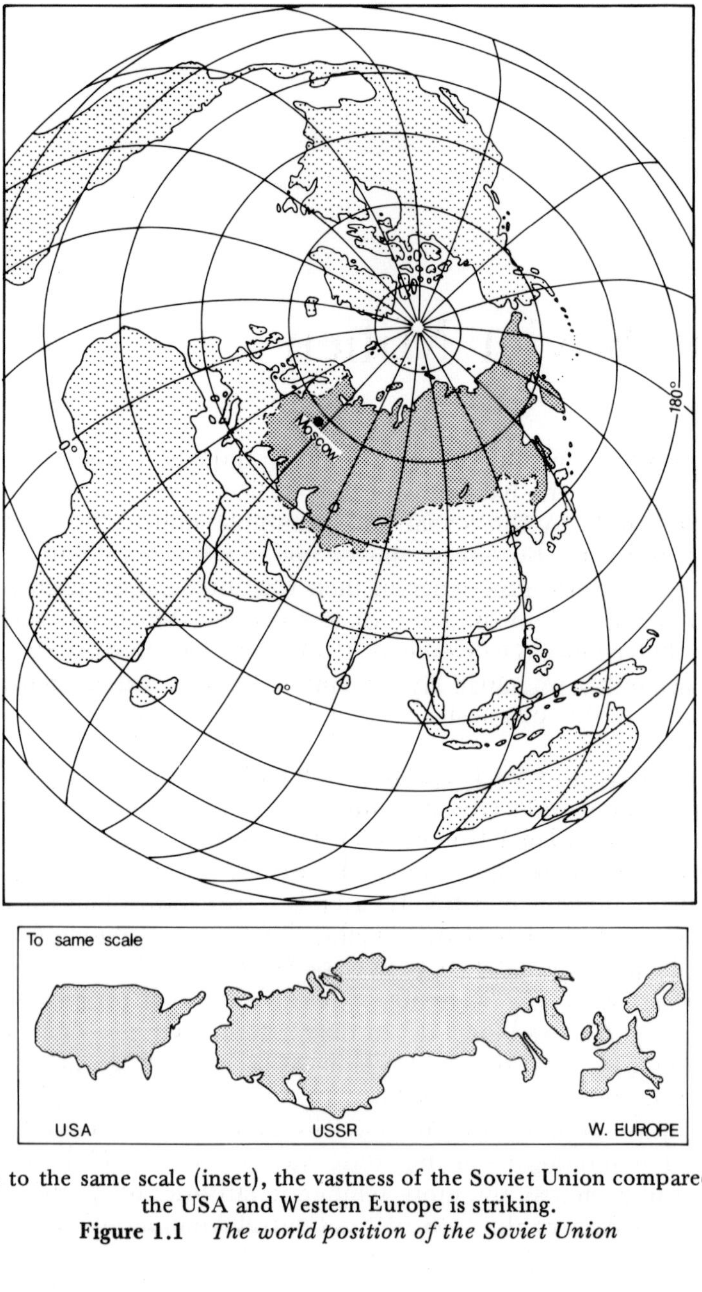

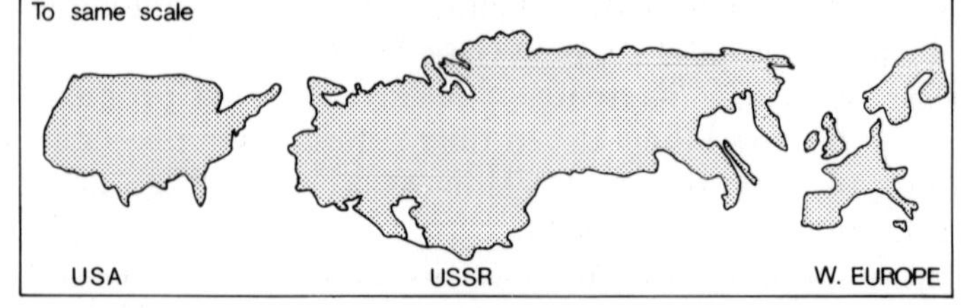

Drawn to the same scale (inset), the vastness of the Soviet Union compared with the USA and Western Europe is striking.

Figure 1.1 *The world position of the Soviet Union*

Stretching across most of northern Eurasia, and comprising some 40 per cent of the Asian land-mass, the Soviet Union holds a commanding position in the great tripartite continental 'world island' – Europe, Asia and Africa – which some students, like Sir Halford Mackinder, have used to create tantalising generalisations about the political and strategic potential of the Russian lands. Covering the great interior heartland, the Soviet Union can act from interior lines of communication on a range of fronts from Europe through the Middle East and Turkestan to Manchuria and the north-west Pacific island arcs, besides holding a long coastline in the now militarily important Arctic basin. Although we have come to accept readily the truly continental nature of the Soviet Union, it should not be overlooked that it has an important coastal element, and although certainly over three-quarters of Soviet territory lie more than 400km (250 miles) from the sea, of its total territorial boundary length of 59,500km (37,000 miles), some 43,500km (27,000 miles) are coastal. This gives the longest coastline of any of the world's countries, but much lies in high latitudes with a long frozen winter period that discourages maritime activity and, consequently, an important element in its political geography has been a search for control of warm-water ports beyond its own coasts.

In the geographical problems of the Soviet Union, the high latitudinal position of much of its territory becomes particularly significant, especially when making comparisons with the USA. Only the Ukraine, Transcaucasia and the Soviet Central Asian republics are in a latitudinal position similar to the north of the USA; even the most southerly Soviet territory is comparable in latitude only with San Francisco and New York. Moscow is further north than any of the major cities of Canada; Kuybyshev is at about the same latitude as Vancouver; and Leningrad is a little further south than Anchorage in Alaska. Even in European terms, some interesting comparisons emerge: Leningrad is roughly on the same latitude as the Shetland Islands, while Moscow is as far north as Edinburgh. This northerly location relative to the main economic and population foci in other continents is exacerbated by the broad outline of relief, with the great plains that dominate its relief open to the Arctic coast, allowing easy ingress of cold polar air. In the south, however, almost like a giant amphitheatre, relief is mountainous, rising up into the high Pamir and Himalayan ranges beyond Soviet territory. The effect is to prevent entry of warm, humid tropical air, so strengthening the negative temperature anomaly over most of northern Asia. The huge continental extent allows rapid warming of the land in summer and cooling in winter. Consequently, the Soviet Union has some of the greatest annual temperature ranges on the earth's surface. Prevention of the ingress of moist air by these mountains also makes the Asian interior notably dry.

With such massive plains, relief plays a modest role in regional differentiation; differences arise far more from the association of climate, soil and vegetation. The vast plains of European Russia merge southeastwards into those around the Caspian Sea that extend away into the Turanian plains of the Aral Sea basin, which again merge imperceptibly in the north into the vast flatness of the West Siberian lowland. Between the plains of European Russia and the even more featureless ones of Western Siberia lie the low ranges of the Ural mountains, whose low central section provides easy passage across the conventional divide between Europe and Asia. The dimensions of the plains are so great that any impression of variation in elevation as shown on atlas maps is completely lost in visual terms to the traveller across them. Although the major elements of the relief can be seen to influence such primary features as the pattern of drainage, it is really the micro-relief, the highly localised, small forms (like steep bluffs along the right banks of rivers), which has the greatest meaning for man (Figure 1.2).

From an atlas map, east of the Yenisey relief appears more varied in the broad Central Siberian Plateau and the vast undulating surface has in places been deeply incised by rivers. To travellers along these valleys there is an impression of mountainous or hilly country until the slopes are climbed and the rolling interfluves are reached, but this upland dips eastwards into broad plains in the Lena basin. East of the Lena river and in Southern Siberia lies part of the mountainous amphitheatre. Until recently this was almost unknown country and the mapping of the mountains of north-east Siberia (highest point, Gora Pobeda, 3,147m, 10,325ft) only properly begun in the 1920s was not completed until after the Second World War. Rough dissected mountainous country forms much of southern Siberia, notably around Lake Baykal (the world's deepest), where some of the higher peaks exceed 2,000m (6,562ft), and on the east the Yablonovyy and Stanovoy ranges comprise exceedingly rough terrain. West of Lake Baykal, in the Sayan and Altay mountains, the highest surfaces lie well over 3,000m (9,842ft) (Mt Belukha, 4,506m, 14,783ft). South-west of the Altay the mountain wall is broken by the strategically significant Dzungarian Gates leading into Chinese Turkestan. Further south, the mountain wall rises to the massive barrier of the Tyan Shan, where large areas lie above 4,000m (13,123ft), with the maximum elevation of 7,439m (24,406ft) in Pik Pobedy. In the panhandle of Soviet territory in the Tadzhik Republic, the high Pamir, much a great dissected plateau, reaches

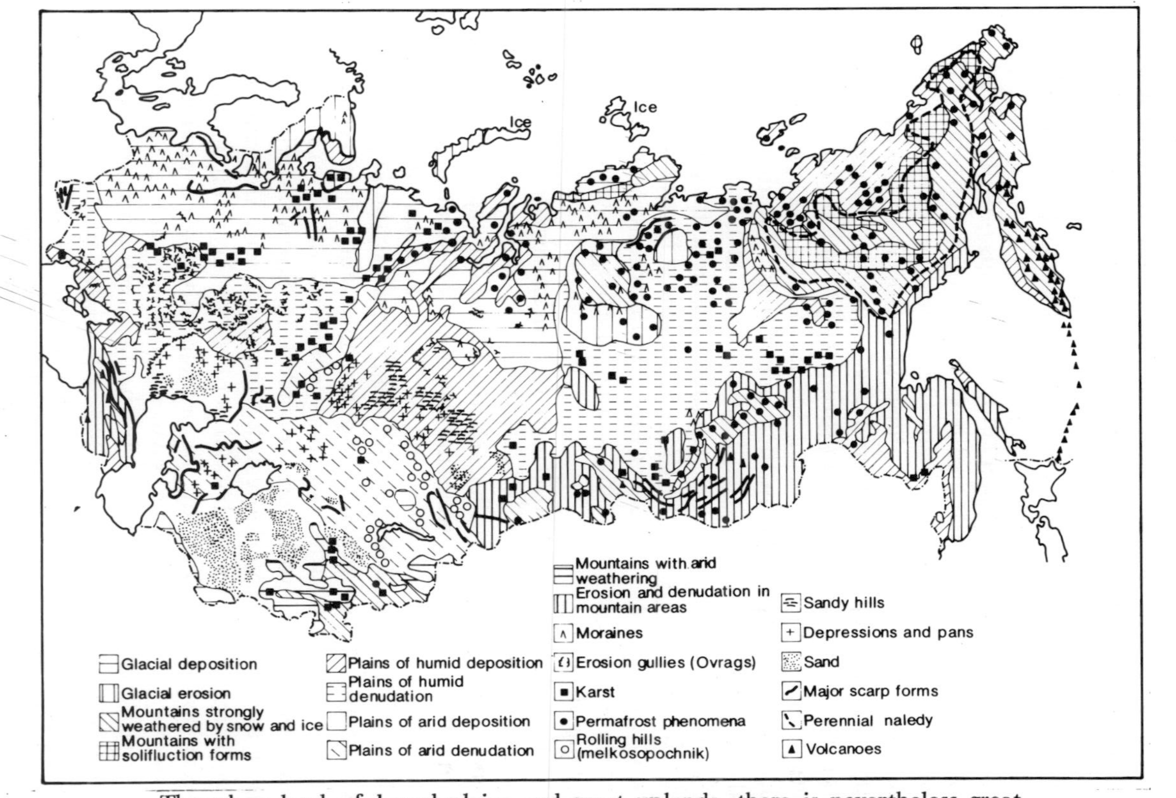

Though a land of broad plains and great uplands, there is nevertheless great diversity in the terrain as reflected by this map.

Source: *Atlas SSSR* (Moscow, 1969).

Figure 1.2 *Terrain types*

7,495m (24,590ft) in Pik Kommunizma, the highest point in the USSR. Westwards towards the Caspian Sea, the country is lower but nevertheless rough. The Caucasian isthmus is really mountain country, comprising the ranges of the Great and Little Caucasus, where maximum heights exceed 5,000m (16,400ft). In the extreme south-west of the country, in the Ukraine, Soviet territory extends into the low ranges of the Forest Carpathians, pleasant rolling mountains crossed by easy passes.

The great rivers of the Russian lands are perhaps one of their most distinctive features. Some rise in the southern mountains, but others gradually gather their strength in the plains as tributaries flow together. Traditionally the rivers have been routeways, with movement up one stream and across a portage to another. Frozen for long periods in the winter, they thaw in spring to spread inundation across immense tracts, and those flowing to the Arctic seas are a special hazard, since they thaw first in their upper reaches in the south and pour down on to the still frozen lower reaches in the north. In the spring thaw, wherever ice floes jam the river bed, they pond back water until the pressure breaks them down and a flood wave sweeps all before it down the valley. Yet however destructive they may be, these rivers are also a promising source of energy, still only superficially tapped.

No reader of a Russian novel can fail to appreciate the role the seasons play in everyday life. Of them all, winter is the most notorious – long, harsh and hazardous. The short capricious spring passes quickly into the welcome warmth of summer, with its thunderstorms, before a short but brilliant autumn heralds once again dreaded 'general winter'. Cold is nevertheless the key word to all the climatic zones and winter gets colder and longer generally eastwards and northwards, so that some of the lowest winter temperatures in the world are recorded in north-eastern Siberia around Oymyakon and Verkhoyansk. Associated with this cold is the likely occurrence over more than 40 per cent of the country, notably in Siberia east of the Yenisey, of permanently frozen ground (so-called *permafrost*), presenting civil engineering and all forms of building with a serious challenge. In the far north permafrost occurs as a complete cover, but towards the south, islands of thawed ground appear; further south still, these grow in size until towards the southern limit only patches of permafrost remain, but it is liable to develop in free ground should the thermal balance be upset. The frozen layer may be many metres thick, so that subterranean water under pressure wells up through cracks to form massive blisters of earth at the surface (*pingoes* or *bugulnyakhi*) that leave the surface pitted with little round ponds when they collapse. Everywhere, permafrost gives rise to asymmetrical valley shapes and surface distortions.

The high degree of continentality is also reflected in the generally low levels of precipitation, much of which occurs in the form of snow, even in Central Asia. In the south, low precipitation is accompanied by high evaporation, where most rainfall comes in thundery summer downpours on parched ground, with much loss by run-off. In the Turanian Basin meagre and unreliable precipitation generates critical moisture deficiency and consequently causes serious aridity. In the north, in contrast, low precipitation (notably as snow) is matched by low evaporation, but the nature of the relief and soil allows a lot of standing water through poor drainage. Consequently the north has in general a surfeit of water, with large tracts of bog and swamp. Continentality is also reflected in summer by the wide diurnal variation in temperature: cool mornings rapidly warm up into early afternoon, when the blue skies cloud over and a late afternoon thunderstorm is not uncommon, leaving a bright, fresh evening. Once the sun sets, the night becomes quite chill, even in the height of summer. Open from the north, once winter comes, cold air rapidly settles across the country. The vast plains are frequently swept by bitterly cold blizzards that make outdoor life hazardous for man and beast, but between these furies come long periods of intensely cold, still air, when blue skies and sunshine bring a feeling of well-being. Unfortunately, because of the penetration of most maritime air in winter into European Russia, the westernmost parts are less pleasant than deeper into Siberia, where the strong development of high pressure gives long runs of still, anticyclonic air. Spring usually comes quickly, sometimes with a föhn-like effect, but the thaw makes the countryside wet and miry (the *rasputitsa*), when movement is difficult and unpleasant. To a person accustomed to living under a more maritime regime, the strength and regularity of the annual continental rhythm is uncanny: there is a truth in the saying that the Russian lands have climate but no weather!

The great latitudinal extent of the major plains results in considerable difference in climate between their northern and southern parts, but over correspondingly large tracts there is relatively little change in the underlying rock formations. Soviet soil scientists accept the ideas of Dokuchayev that climate and vegetation are more significant agents in soil formation than the underlying lithology. Indeed, the major soil types, so influential in the agricultural scene, are closely related spatially to the climatic and vegetational associations of the main natural zones like the tundra, tayga and steppe, etc., and regional description becomes essentially related to these remarkably regular tripartite associations.

The major landscape zones, as Soviet geographers term them, represent a sequent transition from cold polar conditions through warmer and moister conditions to the truly arid interior continental conditions of the far south. The regular latitudinal sequence is only distorted through variations in elevation, as in the Ural mountains, where northerly associations are carried far south. Local conditions on the Soviet periphery, as in humid western Transcaucasia or the monsoonal belt of the Siberian Far East, are exceptions to the general symmetry. These diverse landscape zones, so brilliantly described by L. S. Berg, become the key elements in everyday life. Although definitive lines are drawn between these zones on atlas maps, in reality the transition is gradual. Historically, some belts, like the steppe, have provided great routeways, while others, like the wet and marshy northern forests, have been places of retreat for persecuted and harassed groups.

1.1 **The major landscape zones** (see Figure 1.3)

The most northerly Soviet territory, the offshore Arctic islands and the Taymyr Peninsula, is really polar desert, with poor moss and lichen vegetation or small plants in the most sheltered parts. The fauna is even sparser than in the tundra to the south and the flora also much poorer. There is a small year-round population in meteorological and scientific stations. Large parts of the arctic islands are ice-covered, and everywhere there is a deep and continuous layer of permafrost.

The tundra

To the south most of the Soviet Arctic shoreline is marked by a broad belt of tundra that corresponds to the extent of subarctic climate. Here again permafrost forms a deep and continuous cover. Although precipitation is low, most coming in the winter snow, the frozen sub-soil and the featureless relief over great areas impede easy drainage, so saturated, water-logged ground predominates. The soil is thin, wet, raw and peaty, and only on the better-drained patches does poor podzol occur. Although there are extensive spreads of peat, because of poor decomposition conditions for organic material, it is seldom of any great depth. Wet clays form 'spotty tundra', with clumps of vegetation between barren patches. Very large areas are covered by bogs, but the low-lying country is generally drier towards the east.

The tundra is a treeless landscape, for trees cannot survive in the wet, open, windswept ground. At best there are hardy woody ever-

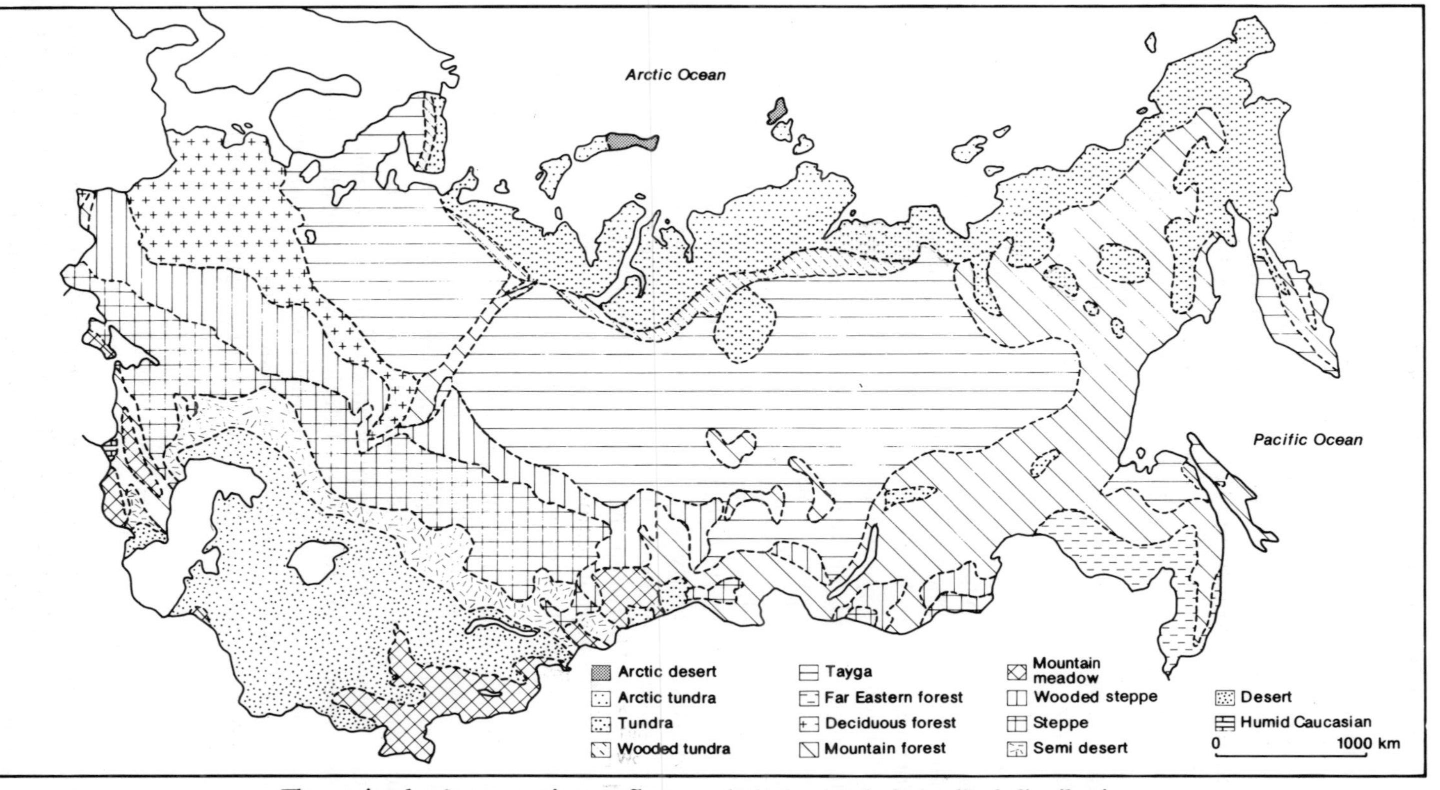

The major landscape regions reflect a relatively simple latitudinal distribution.
Source: modified from *Atlas SSSR* (Moscow, 1969).

Figure 1.3 *Major natural landscapes*

green perennials that grow a few centimetres above the ground, but there is much moss and lichen, and in the spring brightly flowering small plants form swards of colour. Strong winds inhibit upward growth and permafrost hinders establishment of deep roots. However, on travelling south, the landscape changes to 'wooded tundra', first marked only by a few miserable stunted trees like low bushes in the more sheltered localities. These become more common and larger towards the south, until real clumps of small trees as well as scattered individuals appear. On the permafrost, however, the trees are often tilted at crazy angles, with the best and thickest clumps found on the better-drained sandier patches.

Winter in the tundra is bitterly cold, with the landscape grey and dead beneath the endless leaden sky. So long as the air remains still, the cold is bearable, but when blizzards bring snow, life in the open is almost impossible. Winds sweep the snow into hollows or against slopes, leaving most ground bare to the deep penetration of frost. Summer is short and relatively warm, an impression strengthened by the long days, unlike winter when there is hardly any daylight. The summer warmth is sufficient to thaw only the surface of the permafrost that exerts a cooling influence on the air near the ground. Unfortunately, along the coasts and the lower reaches of the great rivers, misty and cloudy weather sometimes prevails for weeks on end and makes conditions unpleasantly raw. The most continental conditions occur between the Yenisey and the Lena rivers, with the greatest annual range of temperature. Spring temperatures rise slowly and autumn is usually shorter than spring.

The long dark winter in the arctic desert and tundra leaves them almost without animals. Only the odd polar bear or seal ventures along the coast in winter, but in summer reindeer penetrate northwards from the forest fringe, and immense numbers of migrating aquatic birds fly in from the south, followed by predators such as foxes and the wolf, which also prey on ermines, ferrets and lemmings that come north from the forest. Life is, however, made unpleasant by gnats and mosquitoes. Apart from its wealth in fish and animal fur, the tundra has little attraction to man, with a few native hunters and poor reindeer herders along its southern fringe as well as some coastal dwellers.

The tayga

Moving southwards the trees grow bigger and the clumps turn into stands of woodland and eventually into great forests. This is the boreal coniferous forest, the tayga, a broad belt that extends across Russia's middle latitudes, where the predominant podzol-type soils, poor and acid, are distinguished by a thin raw layer of peaty humus derived from the conifers. These are difficult soils to farm, but towards the south, where leaching is less, they can be useful for cultivation. The tayga forests are, however, broken by extensive tracts of bog and swamp, most commonly north of 60°N, while broad strips of rank meadow occur along the rivers. The Lena basin around Yakutsk has quite good and rather limey soils, while generally east of the Yenisey conditions are drier and bog and swamp less. In contrast, the West Siberian lowland is largely swamp, which shows many signs of having encroached upon forest in recent historical times.

The unmanaged primordial coniferous forest of the tayga contains about a third of the total world forest in a belt about 4,800km (3,000 miles) long by some 950km (600 miles) broad. Where the great Siberian rivers carry warm water to the Arctic Ocean, tayga extends north along them into the otherwise tundra belt. With the vegetative period lasting seldom more than four months, the number of species is limited to conifers with low transpiration and able to withstand strong, cold winds. As the winter snow is less easily blown away in the forest than in the tundra, it lies deeper and provides an insulating layer for small plants of the forest floor. The species in the tayga occur in extensive stands of the same type, a great help in their economic exploitation, though the trees are generally small. Wherever conditions are suitable, they grow close together, so that the forest floor is covered by an acid layer of needles and cones, with fallen, decaying trees. Much of the northern edge of the tayga comprises stunted trees of little economic value, just as there is everywhere a great deal of overripe timber in these natural stands. The forest, by impeding drainage on the low interfluves of the West Siberian lowlands, has assisted the encroachment by raised bog, while in many parts the tilted trees of 'drunken forest' caused by permafrost pose a problem for exploitation. This is a gloomy, silent, forbidding forest; and where the stands are thickest, to venture into them, it is said, is to know the meaning of fear. In Eastern Siberia, where the country is drier, vast forest fires burn for long periods in summer, filling the air with acrid wood smoke, and leaving the country bare to regenerate naturally. Consequently, forest in varying stages of recolonisation is found.

Spruce and fir dominate in the European tayga, with pine on sandy patches (*bory*), but despite the poor soils and the troubles of farming them, much land has been cleared for cultivation. Spruce and larch typify the northern Ural, while pine, birch and larch are common in the south, where the forested mountains stand above the surrounding steppe. In Western Siberia spruce is common in the north, where it does well on permafrost because of its superficial root system, and larch is also common, though bog reaches its greatest extent in this part of the tayga, but the southern part of Western Siberia has characteristically thick stands of fir. Drier Eastern Siberia is mostly larch forest (including the valuable decay-resistant Daurian larch), but in Transbaykalia rich pine forests cover the valleys. With better drainage in the more mixed relief, and lower precipitation, bog is less common, but the drier conditions substantially raise the fire risk. The Lena–Vilyuy lowlands near Yakutsk, with their carbonate soils, have extensive grassland. In Southern Siberia the varied relief carries forest on the uplands far south into the grassland belt and forested ranges stand above grassy valleys and basins. Kamchatka has mostly thin, park-like birch forest.

The moderately long summers of the tayga are sufficient to allow trees to grow and fruit, but winter temperatures are in many places much lower than for places on the Arctic coast, especially in valleys and basins where cold air drainage takes place. Nevertheless, with long stretches of sunny anticyclonic weather and remarkably still air, despite the intense cold, it is not unpleasant in winter and the air is remarkably healthy and germfree. The most intolerable conditions occur once again in the occasional *buran*, a violent blizzard-like storm, during which it is often impossible to go out of doors for many days on end. Summers are warm but the diurnal range is great, with daytime temperatures of 35°C (95°F) dipping to 5°C (41°F) at night, but the enjoyment of summer is spoilt by plagues of flies and midges. The long summer daylight allows vigorous plant growth, offsetting the short winter days. On the whole, there is less contrast between east and west in this belt in summer than in winter.

Relatively few animals live in the deep tayga forests and their absence makes the gloom all the more oppressive. Around the forest fringes, in clearings and open patches, are found the wolf, lynx, bear, fox and many small furry animals. Of these, the silver fox, ermine and squirrel as well as the Siberian sable first attracted Russian settlers. So many animals were taken that the survivors were driven into the more inaccessible parts of the forest, to which trappers and hunters followed them, and this quest to a large degree generated the Russian exploration of Siberia. Deer have been exterminated in the European tayga, but they still occur in Siberia, mostly elk and maral. The tayga gives temporary home to many birds of passage as well as to ptarmigan, capercaillie and other forest species. Vipers and harmless reptiles occur, but the main hindrance to human colonisation comes from the prolific midges and gnats of summer.

The mixed forests

The southern edge of the coniferous forest gradually begins to include more deciduous tress and eventually the broad-leaved trees predominate. This belt of mixed forest, the dominant form in European Russia, lies like a wedge with its apex in Siberia towards the foot of the Altay, though it reappears again with different species in Amuria. It is underlain by modified, less acid podzol and even grey forest soils, while dark soils on the southern edge suggest an invasion of forest into grassland. These forests are lighter and more open than the forbidding tayga. Oak and spruce are typical, while hornbeam is common in the south, but ash, elm and maple also occur. There are patches of swamp, as in the Polesye, but the forest has everywhere been widely cleared for farming, while long occupancy by man has also modified its original form. It is the better soils that make it so attractive to farmers, but long colonisation has drastically reduced the native fauna, otherwise similar to the tayga. Within this belt the last European bison are found in reservations.

The European mixed forest has long cold winters but very warm summers, though penetration of cyclones into north-west Russia gives a more varied winter than deeper into the continent. Winter in north-west Russia is often broken in November and December by humid mild spells, when the raw air gives rise to many respiratory complaints. The moister winter here is gloomy and overcast, more oppressive and trying than the brighter but colder anticyclonic weather experienced in Siberia. About a third of the precipitation comes as snow brought in moisty cyclones from the Atlantic. From late December the rivers freeze and remain so until March or even into April. After a wet spring following the thaw, summer quickly warms up, but July and August are often cloudy and overcast, oppressive with thundery downpours that give about a third of the total precipitation, though these at least settle the dust blown in from the southern steppes and often associated with disease. With a wide diurnal range of temperature, summer frosts are not unknown. Autumn is bright and pleasant, but it passes more quickly into the grip of winter than in western and central Europe.

The mixed, broad-leaved forests of the Amur basin have a monsoonal climate. The summer is moist and warm, tempered by an inflow of Pacific maritime air, with monsoonal rains brought by winds from the sea from May to September, with a maximum in July. Precipitation, however, declines rapidly away from the coast. Winter is cold, like in the Siberian interior.

The steppe

The forest slowly becomes more open and patches of grassland grow in extent as one travels south. The transition from forest to grassland, a park-like landscape with clumps of trees amid a grassy sward, forms the 'wooded steppe', though almost none of the natural form now remains, for here man has taken more land than elsewhere into cultivation. The true steppe is where grass alone dominates the landscape, on the northern side as a complete sward, but further south, as the country becomes drier, breaking into clumps and grassy patches with bare clay between as the semi-desert is approached. The steppelands have been the historical routeway for migrant people moving from Asia to Europe. Under pressure from fierce steppe nomads, early Russian settlement retreated into the forest, but recolonisation began with Russian domination in the eighteenth century, when much of the European steppe was named 'New Russia'. The natural home of grasses, much of the better steppe, especially in European Russia, has been taken into grain cultivation since the early nineteenth century. The northern edge of the steppe has quite fertile modified forest soils, while even the poorer chestnut and brown soils can be made fertile by careful irrigation. The steppelands are, however, characterised by large spreads of the remarkably fertile black earth (*chernozem*), with a loose crumbly surface layer rich in humus, deep chocolate to black in colour, and well drained, especially where it overlies *loess*. Because of the intense evaporation in summer, salts are easily drawn to the surface, particularly in depressions with a high water table, so that where irrigation is used, great care must be taken to avoid surface salt formation, not always the case in the past.

Winters in the steppe are cold, with strong winds sweeping the open surface, blowing the sparse snow away and exposing the earth to intense frost penetration. The friable soil is thus easily reduced to dust, the raw material for duststorms which carry away valuable topsoil and winter-sown crops. Spring comes slowly, but the thaw turns the soil into a muddy quagmire. Rainfall in late spring brings a carpet of brightly flowering plants, though young crops may be damaged by run-off in thundery showers. Summer becomes progressively drier as the landscape turns a parched brown and duststorms occur more frequently, but particularly unwelcome is the withering dry *sukhovey* wind. Diurnal range is great: clear, cool bright mornings turn to stuffy, cloudy and oppressively dry afternoons, broken commonly by a thunderstorm that clears and cools the air. The clear air of autumn gives bright and distant horizons that make the emptiness of the steppe look even vaster. In general, the steppe in

European Russia is cooler in summer but warmer in winter than that in Asiatic Russia. Precipitation is at a maximum in spring, but is nevertheless low, decreasing eastwards and southwards. With between 200–400mm (8–16 inches) per annum, it is poor for agriculture, especially because of high moisture losses through high evaporation in summer. With much rain coming in thundery showers on to parched ground, run-off losses are excessive and gullying a serious problem throughout the steppe. One of the greatest difficulties arises, however, from the wide variation in precipitation from year to year; such is the unreliability over many parts of the Siberian steppe that at least one year in five is a complete disaster for grain crops.

The natural fauna of the steppe would be antelopes, deer and wild horses, but excessive hunting has exterminated most species. Ground-nesting birds are, however, typical, but they have suffered seriously as the steppe has been ploughed and harvested. Small rodents are now the most common element, frequently reaching pest proportions, doing much damage to crops. From time to time there are also incursions from grasshoppers and even locusts.

The desert

Over most of Soviet Central Asia low precipitation and high evaporation mean intense aridity. Consequently the characteristic landscape is semi-desert or even true desert, with some poor steppe, though remarkably luxuriant vegetation arises wherever rivers from the mountains provide water. Salts drawn to the surface by the intense evaporation have resulted in great barren tracts of several types of salt pans, while the desert is dominantly clayey or sandy, influencing soils. On the southern fringe along the mountain foot, broad spreads of soils, apparently of loessic origin, when properly irrigated and farmed, can be exceptionally fertile, like the silt along the rivers. The plants of the desertlands are all adapted to resist loss of moisture and to flower and fruit quickly in brief moist periods. A rare rainstorm transforms the drab landscape into a sea of brilliant flowers. Trees are absent, except where there is water in oases, which can be extraordinarily lush and fertile, like the *tugai* jungle along the Amu-Darya, and in the Semirechye. In the moister desert the most common tree is the dead-looking *saksaul* growing in thickets that give almost no protection from the sun and by excluding any wind are hotter inside than in open country. The fauna is well adapted to resist drought and heat, so that many species are dormant for long periods. Small rodents and hares are found along with their predators like foxes and wild cats, but antelopes, gazelles and the wild ass are now rare. Boars, jackals and even tigers occur in the *tugai*. Birds are few but reptiles and insects common.

Over this area, precipitation is generally below 250mm (10 inches) annually, but it is also very unreliable and the position of the 250mm-isohyet shifts considerably from year to year. Summers are hot and dry, comparable with those of North Africa. The brief autumn gives way, however, to a remarkably cold winter, when rivers freeze and cold winds from Siberia sweep relentlessly across the open landscape. Winter passes into a short spring, when most of the scanty precipitation falls. For man the intense heat of summer is reasonably bearable since the air is extremely dry, though occasional scorching winds like the *gamsil* make conditions trying.

The Transcaucasian lowlands

The lowlands of Transcaucasia form a distinctive natural landscape. On the west the Kolkhid lowland is described as 'humid sub-tropical' by Russian geographers, where hot summers and mild winters allow plant growth the whole year, while the heavy precipitation is fairly evenly distributed the year round. Soils are generally fertile, including some good red soils, and the vegetation is luxuriant. The area can produce crops like citrus fruits and tea. The eastern lowland around Lenkoran is drier, with a marked dry season, and much poor steppe and semi-desert. Winter here is cold, often extreme. Nevertheless, where there is moisture, vegetation is lush, but the evergreens of the west are absent.

The mountains

Elevation of the earth's surface in mountains may be sufficient to create the conditions of particular landscape zones in somewhat modified form far to the south of where they might be expected to occur at or near sea level. The effect of altitude is particularly well seen in the mountains of Soviet Central Asia – at their foot and lowest slopes, arid spreads of piedmont loess can be converted into rich farming country once irrigated; in the foothills the desert quickly gives way to steppe with a modest rise in elevation. The steppe becomes more luxuriant as the mountains are climbed until at about 600–800m (2,000–2,700ft) in the Tyan-Shan, for example, unirrigated land becomes more common than irrigated; by 1,000m (3,500ft) on the wetter slopes there is already forest, both coniferous and deciduous, and somewhere around 2,400m (8,000ft) alpine meadow appears. Mountain sheep and goats are common and there is also the ibex and snow leopard.

In the Great Caucasus the altitudinal succession is similar, but in the drier eastern ranges steppe conditions prevail to a greater altitude than in the west, which is appreciably moister. In the mountains the alpine zone begins at between 1,800–2,100m (6,000–7,000ft). Caucasian ibex, chamois and snow pheasant are typical of the fauna. The Little Caucasus and Armenian Plateau consist of an extensive steppe, though summers are cooler than in the true steppe and there is a deep winter snow cover. In Armenia the fauna and flora comprise many elements from Asia Minor. In the sheltered valleys of Armenia vegetation degenerates into semi-desert.

The relatively low Ural mountains carry northern conditions far south, well seen in the deflection of the boundaries of the major landscape zones on the atlas map. The highest surfaces push the tundra well south and the forest belt extends in tree-covered ranges into the steppe that laps round the mountain foot. In the central forested Ural the alpine zone is reached only by some small and particularly high summits, while few parts of the southern Ural rise above the forest zone. In the mountains of southern Siberia, notably the Altay and Sayan ranges as well as the ranges of the Baykal lands, forested mountains are separated by steppe-like basins into which strong elements of Mongolian flora and fauna have penetrated. The effect of the mountains of northeastern Siberia is a massive extension southwards of the poor tayga giving way to huge areas of barren tundra conditions around 600–700m (2,000–2,300ft). East of the Kolyma river tundra predominates over forest and there are immense tracts of bare, desolate mountains that give little support to man or beast.

1.2 Environment and the Soviet milieu

In this brief review of the physical environment two recurring themes have appeared. First, there is the immensity of distance, to form, in a sense, the embarrassment of space, and second, the ever-present harshness of the physical environment. These are the inescapable facets of the social and economic milieu in which all Soviet people live out their everyday lives, while most of these landscape zones are in themselves so vast that the majority of the population spend their whole lives in one such zone, sometimes completely without experience of any others. Historically, the landscape zones like the steppe or the forest have witnessed the evolution of their own distinct social and economic systems. Although figures have been quoted for the area of the Soviet Union and comparisons made with the United Kingdom and the USA, it is perhaps more telling to make comparisons of actual distances between places and the size of features of the landscape. For example, one thinks of London and Edinburgh as being at opposite ends of Britain, yet such a distance (roughly from New York to Cleveland in the USA) is only as far as from Moscow to Leningrad, which appears so modest on the map of the USSR, while Moscow is further from Vladivostok, the other end of the Soviet Union, than London is from New York. Such distances in themselves create problems of communication: how can the bureaucrat in Moscow speak by phone to his opposite number in Vladivostok during either's working hours across a time gap of a working day? Or we may consider some physical features in terms of dimensions we understand. To start, the Caspian Sea appears as a lake, albeit extremely large, on our atlas map, but in terms of area, it could swallow the United Kingdom, though it is not quite as large as Montana, USA. It is then not surprising that the ferry boats from Caucasia to the Central Asian shore are larger than those running across the southern North Sea from Britain to the European Continent. Lake Baykal appears a modest feature buried deep in Siberia and we may wonder why it seems to play such a vital role in the life of that part of the USSR, but it falls into perspective better when we realise that from one end to the other is further than from, in the United Kingdom, Bristol to Edinburgh (say, Washington, DC, to Boston in the USA). Sea-going ships of over 8,000 tons may sail up the Yenisey to Igarka – further than up the Rhine from the Hook of Holland to Basel – and the lowest reaches of the river Ob are considerably wider in parts than the English Channel. All economic geographies of the Soviet Union describe the Ural industrial area, which in our mind's eye we doubtless compare with Western European industrial areas like the German Ruhr or the North-east of England, but in reality this region covers an area equal to England and Wales together. Even the line of industrial towns forming the Kuzbass extends as far as from London to Carlisle or from Chicago to St Louis! Distance consequently becomes a critical factor in considering resource evaluation. While gross totals for reserves and production of minerals or the manufacture of goods or even for such important resources as population may be formidable, bringing them together in effective interaction raises all sorts of difficulties. Even to reach a given level of development or provide a given service over such vast space usually involves a disproportionately larger amount of resources than to achieve the same in a smaller, more compact setting. This is reflected, for example, in the greater average length of haul for most goods like coal or building materials in the Soviet Union compared with, say, Western Europe. Coal moves on average ten times as far by rail in the Soviet Union as it does in the United Kingdom. Another transport analogy also illustrates the problem: to run one train a day each way between London and, say, Aberdeen, a major axial distance in the United Kingdom, could be done with just two sets of coaches. Over the comparable journey in the Soviet Union, between Moscow and Vladivostok, the same level of service would require something in the order of sixteen sets of coaches. Even to build a branchline may also swallow massive resources, so that the Tayshet–Ust Kut branchline in Siberia, as far as from London to Dundee, even as a single track route required at least 86,000 tonnes of steel for the rails alone. The present construction of the Baykal–Amur trunkline, some 3,200km (almost 2,000 miles) long, will need at least half a million tonnes of steel. Another example of the role of distance can be drawn from the energy sector. Although the daily pattern of peak loads of electricity consumption is the same throughout the Soviet Union, the distance between Moscow and Western Siberia means at least a two-hour difference in occurrence, which could be used to advantage to transmit current between the two regions to even out the load. It is thus easy to understand why Soviet technologists have devoted themselves notably to the problems of extremely high voltage transmission, the only way of successfully transmitting current over long distances. Such a problem is not faced in Western Europe, either on a national or an international scale.

With some six or more hours difference between European Russia and the Far East of Siberia, problems of communication are clearly more intense than the four-hour difference in the USA and quite out of scale to the one-hour difference over part of the year in the European Economic Community. Time–distance relationships clearly have a special meaning, and it is perhaps not surprising that the Soviet Union is notably air-minded, with an acute interest in high-speed communication. This time–distance problem becomes a significant factor in the shortcomings of running a centrally planned economy. Computer handling of data doubtless provides an even greater attraction to the managers of the Soviet economy than to those running more complex but spatially more compact economies in Western and Central Europe. The observant traveller will nevertheless wonder at how successful the Soviet authorities have been in achieving standardisation across such an immense and diverse territory.

The second aspect, that of climatic harshness and its influence on everyday life, is reflected widely in transport, the construction industry and settlement. The influence of permafrost across more than 40 per cent of the country is a distinctive feature of the Soviet scene likened only in northernmost Canada. In making comparisons between the performance of the Soviet Union and other countries, it is clear that the greatest difficulties arising from the adversity of climate concern the farmer. When we consider the areas unsuited to farming through a surfeit or deficit of water or through an unsuitable thermal regime to allow the fruiting and ripening of crops, we see how the remaining area bears a close resemblance to the over-all general pattern of settlement. The Soviet *ecumene* remains a broad triangle with its base between the Baltic and the Black Sea and its apex in Western Siberia: this is the area suited to sedentary farming of some sort or other. Within this triangle the narrow east–west strip between the southern forest and the drier steppe remains the most thickly settled, being the area most suited to arable farming. Only in the rather special conditions of the oasis lands of Central Asia and in the most fertile parts of Transcaucasia do similar population clusters occur.

The Soviet Union is an interesting example of efforts to overcome the environmental constraints on society and the economy. The older school of geographers who accepted environmental determinism found many of their examples in Russia. Marxist-Leninist philosophy has, however, accepted cultural determinism in which ruthless application of modern technology can rectify nature's errors and surmount its limitations. In Stalin's day grandiose schemes under the general heading of the 'transformation of nature' were proposed (some needing gigantic constructional work), like a vast scheme for shelter belts to ameliorate the climate of the European steppe, the fantastic Davydov Plan to collect the waters of the Ob–Irtysh rivers in a shallow but immense dam in the West Siberian lowlands and then divert them through the Turgay Depression to 'top up' the falling levels of the Aral and Caspian Seas, or a macabre scheme to dam the Bering Strait to cut off the cold Oya-Siwo current and so warm up the Soviet Far East. The ploughing up of the Virgin Lands from Northern Caucasia across Kazakhstan to the foot of the Altay, inspired by the botanist and geneticist, Trofim Lysenko, in its partial failure showed only too forcefully the delicate balance of nature, whose upset without adequate long-term scientific investigation could lead to disaster. Even the apparently reasonable idea of a cascade of hydro-electric barrages along the Volga has brought unwanted and unexpected side-effects. By slowing the river's flow, the frozen period may last for anything up to ten days longer in many reaches; acute siltation accentuated by natural tide-like seiches in the vast shallow reservoirs behind the barrages threatens their long-term existence; the choppiness of the water in these reservoirs makes the use of traditional shallow-draught river vessels dangerous; by slowing the flow of water and reducing the inflow into the Caspian, the input of nutrients is reduced with sad consequences on the fish stocks, while many fish that formerly spawned in the river have had their life-cycle upset by a change in the thermal regime through altering the natural current. It is clear that qualification is needed for the words of the popular Soviet anthem that claims 'man walks as master of his immeasurable fatherland'.

Conservation and pollution problems in the Soviet setting

Although the arrogance of the Stalinist period towards the 'transformation of nature' has passed, recognition even so of the need for conservation has been only reluctantly accepted. Perhaps in such a vast and modestly peopled country, a feeling that the bounty of nature is inexhaustible is not surprising, but there is also the important attitude of the Soviet arrangement of priorities. Western observers have recorded numerous instances where economic expediency has been given preference over quite pressing conservation needs, despite a growing clamour from Soviet scientists and planners for an enhanced status for environmental considerations.

A clear difficulty arises from a reluctance of many Soviet authorities to surrender an entrenched view that many resources (notably, for example, water) are limitless and will be constantly replaced however they may be abused. One of the most notable cases in this category has been the struggle to prevent pollution of Lake Baykal by industrial effluent to preserve its unique fauna and flora. Although purification plants have been built in many places for industrial effluent, there has often been a long delay in getting them used, while some industrial undertakings have bypassed them where they have been a hindrance to achieving plan targets. Only one-fifth of industrial effluent currently undergoes purification. Another water-pollution hazard has been from oil around offshore workings in the Caspian Sea and other disasters have been recorded from breaks in oil pipelines and even from storage of heavy oils in surface 'ponds' through inefficient production planning. A common problem has been discharge of domestic effluent into rivers through the inability of many urban sewage systems to cope with rapid population growth. Public water supply has clearly been at risk in many places and it is perhaps surprising that the Soviet public health record appears to be as good as it is. The Odessa cholera outbreak in the 1970s was nevertheless one incident revealed to the West; other rumoured incidents have also occurred, including the mysterious anthrax outbreak in the Ural region in 1979.

The same inexhaustible quality is also ascribed by some Soviet planners to air. Many towns have sought the simple solution to noxious emissions from plants by removing the offenders to well outside the towns, while there has been poor control of the period when smoke, etc., may be emitted and even filter devices (especially where these hamper production) removed. Serious air pollution from many metallurgical plants and chemical works in areas like the Donbass and Ural occurs especially in long periods of anticyclonic still-air conditions. Only in the last decade does this appear to have been resolutely tackled. As yet, with a relatively low level of vehicle ownership, pollution from motor-cars and lorries is modest, but the long-term challenge as the upsurge in motor-vehicles takes off does now seem to be recognised. Various rumours have circulated inside and outside the Soviet Union about at least one major atomic disaster, when a large area of the Ural region near Kyshtym was reputedly badly contaminated.

The tribulations of expanding the sown area of the marginal lands of Siberia and Kazakhstan have also brought home to Soviet planners the limits to the belief that 'man marches as master of his immeasurable fatherland', just as it has been realised that land is also not the limitless resource previously taken for granted. It has become apparent that many problems have a deeper cause than the 'capitalist limitations' so long used as an explanation for the failings of Russian agriculture. Certainly one of the most serious challenges has been extensive gullying over vast areas of the European black-earth lands, an inheritance from tsarist times. There has also been reconsideration – as in the West – of the use of pesticides and herbicides on an unlimited scale, though here again economic expediency has often overruled environmental considerations. From the mid-twentieth century the awareness has grown of the need to consider forests as much more significant in the over-all pattern of land use and conservation after decades of ruthless exploitation. Experience along the Volga has shown that major projects like the large dams for hydro-electric generation can have far-reaching consequences for farming through changes in the ground-water table. Equally, the aftermath of opencast mining and the non-recovery of such sites can extend an influence over a large area through such problems as dust generation, change in patterns of drainage or the colonisation of abandoned workings by undesirable flora.

It would be unfair to leave an impression that the Soviet authorities have made no formal provision for conservation of unique elements of the nation's flora and fauna. Much has been done through designation of nature reserves (*zapovedniki*), some of considerable extent, for special flora, fauna or even geological and topographical conditions. Nevertheless, the number of these reserves has fluctuated considerably over time, as has their area, again reflecting the problem of priorities. Great efforts have been made under difficult conditions to preserve threatened species like the Siberian and Caspian tiger, the European bison (*wisent*), the kulan and various species of deer. Similar efforts have been made to protect fish species and also marine animals. In some cases the good work has come too late and certain animals (like Przhevalskiy's horse from Central Asia) can now only be found in zoos.

There does, however, seem now to be a recognition of the need to give greater priority to conserving the natural environment, for which most Soviet citizens have a great respect, as well as to preserving the quality of life in the fullest sense. Perhaps because it is harder to develop a pressure group or lobby in the Soviet Union, in the face of great economic demands, positive action has as yet been weaker than in many Western countries.

1.3 Where to follow up this chapter

The classic description of the major landscape zones remains Berg, L. S., *The Natural Regions of the USSR* (Macmillan, New York, 1950). For Asiatic Russia, see Suslov, S. P., *Physical Geography of Asiatic Russia* (Freeman, San Francisco, 1961). The corresponding volume for European Russia, Dobrynin, B. F., *Fizicheskaya Geografiya SSSR: Evropeyskaya Chast i Kavkaz* (Moscow, 1948), has unfortunately not been translated. A useful recent source is Gvozdetskiy, N. A., Milkov, F. N. and Mikhaylov, N. I., *Fizicheskaya Geografiya SSSR*, 2 vols (Moscow, 1970, 1976), and there is also Gvozdetskiy, N. A. (ed.), *Fiziko-geograficheskaya Rayonirovaniye SSSR* (Moscow, 1968).

Climate is well covered in Borisov, A., *Climates of the USSR* (Oliver & Boyd, Edinburgh, 1965).

On the 'transformation of nature' theme, see Kovda, V. A., *Great Construction Works of Communism and the Remaking of Nature* (Moscow, 1953), and Burke, A. E., 'Influence of Man upon Nature – the Russian View', in *Man's Role in Changing the Face of the Earth*, ed. Thomas, W. C. Jr (Chicago University Press, 1956).

On conservation, a Soviet work is Blagoshkolnov, K. N. *et al.*, *Okhrana Prirody* (Moscow, 1967), but in the English language there is Pryde, P. R., *Conservation in the Soviet Union* (Cambridge University Press, 1972), Singleton F. (ed.), *Environmental Misuse in the Soviet Union* (Praeger, New York, 1976), and Volgyes, I. (ed.), *Environmental Deterioration in the Soviet Union and Eastern Europe* (Praeger, New York, 1976).

2

The Soviet State - its Territorial Origins and its Internal Organisation

The territory of the USSR encompasses the settlement area of the three national groups – the Ukrainians, Byelorussians and Great Russians – that emerged from the eastern branch of the great diaspora of the Slav tribes from their original homeland somewhere between the Carpathians and the Pripyat marshes. In this as yet unexplained fundamental urge to expand and colonise, the Eastern Slavs pushed a wedge of agricultural settlement into sparsely settled forest and steppe far into Siberia, and this process of settlement and expansion is still ongoing in the modern phase of industrial colonisation.

The Russian nation that has taken the leadership of the Eastern Slavs grew in the vast uniform plains of the eastern periphery of Europe, where natural defences were hard to find and yet where distance and immensity gave formidable protection, shielding the early Russian princedoms from being swamped by more dynamic neighbours. The first effective Russian princedom, focused upon Kiev, was organised through outside influence, that of the Eastern Vikings pushing south along the Dnepr to reach the treasures of Byzantium. It was really, however, the powerful but later leadership of the Muscovite princes from their commanding central Mesopotamian position in the plains that set afoot the expansiveness of the Russian state. Its western bounds were set by the stiffening resistance of powerful and better-organised peoples, including Western Slavs like the Poles, and the fluctuations in the position of the westernmost boundary over time reflected the relative politico-military strength of Russia and its neighbours. Southwards, the limits of Russian territory were influenced by the balance of power between the essentially forest-dwelling Slavs and nomads or other forces that held the steppe. So constrained on the west and south, much of the relentless expansiveness of the Eastern Slavs sought an outlet eastwards or northwards into lands sparsely settled by more backward people. In this process Russia had established a foothold by the late seventeenth century on the shores of the northern Pacific and began to turn the emptiness of northern Eurasia into a truly Russian land.

2.1 Territorial growth – the beginnings

The beginnings of the true Russian state date back to about the ninth century AD, when under Eastern Viking or Varangian leadership Slav tribes on the river route from the Baltic to Byzantium were welded into a powerful allegiance centred on Kiev. Under its more successful princes, the so-called Kiev Rus was able to spread its boundaries and draw in many of the small Slav princedoms developed in the basins of main rivers of western Russia. Nevertheless, Kiev was under frequent attack from warlike nomads who every so often spilled across the steppe from inside Asia. One of the most devastating of these, the Pechenegs, besieged Kiev, which was saved by the great Prince Vladimir, who late in the tenth century brought Byzantine Christianity to the Russian lands. The nomad incursions drove many Slavs from the wooded steppe deep into the northern forests, where in the eleventh and twelfth centuries, and again in the Mongol times of the thirteenth and fourteenth centuries, there flourished the golden age of monasticism that was to have an effect on the economy and society of northern Russia until the Revolution of 1917. The north was really dominated in these early times by the great trading and commercial town of Novgorod, but in the twelfth century political influence moved increasingly to Vladimir-Suzdal, forerunner of Moscow.

The thirteenth-century Mongol invasions marked a major formative period in Russian nationhood. After the destructive initial conquest, the Mongols held much of Russia under their influence from their base at Sarai on the Volga, when the 'pax Tatarica' they imposed opened up trade routes between Europe and Asia. To avoid Mongol domination, the westernmost Russian princes sought an alliance with Lithuania, then a powerful state, and recognised the authority of the Roman Church. The eastern princes, however, saw loyal submission to the Mongol khans as the way of protecting the

Russian lands and Orthodoxy. Once Mongol power declined from the mid-fourteenth century, their own power grew, though by this time they had absorbed many Mongol–Tatar influences that gave the princes an ever tighter hold over their subjects. In this period the central Mesopotamian position of Moscow placed it well to expand its authority over other Russian princedoms, especially as between 1328 and 1480 it held the right to collect tribute on behalf of the Tatars, several of whose princes ultimately transferred their allegiance to it. After the fall of Constantinople to the Ottoman Turks in 1453, Moscow exercised its political powers over Kiev's claims to become the centre of the Orthodox Church, the legendary role of 'Third Rome'. Capture of the vitally strategic 'Smolensk gateway' against the stiffening opposition of the Polish–Lithuanian Commonwealth and the subjugation in 1478 of rich commercial Novgorod added to Muscovite power, which now had access to the Baltic and to the route to Siberia.

The sixteenth century was marked by the absorption by Moscow of Tatar princedoms in the Volga basin, and its power was pushed south to include the freebooting Don and Terek Cossacks as vassals. As the territory was consolidated, Russian settlers flowed in to populate a pioneer belt in the black-earth lands, though much of the steppe and the southern gateways to Siberia remained in unfriendly Tatar or Turkish hands. It had to wait until the eighteenth century before Russian control could be pressed to the Black Sea and to the Azov coast as well as into the Crimea. The major move was, however, into Siberia, which had begun with Moscow's incorporation of Perm and Vyatka late in the fifteenth century, though the real annexation began effectively in 1584 with the campaign of the Cossack Yermak. The 'conquest' of Siberia was triggered by the need for furs for trade, since payment of the *yassak* tribute to the Tatars had depleted the fauna of fur-bearing animals in the European forests. Novgorod merchants in the eleventh century had probably been the first Russians to go into Siberia, but a real spread through the northern gateways that the Tatars could not dominate had begun in the early sixteenth century (Figure 2.1). Following the rivers and moving from one basin to another across portages, the trappers and adventurers, both mostly Cossacks, by now loyal subjects of the tsar, moved with remarkable speed across the vast and difficult terrain of this new land. By 1639 the Russians had reached the Pacific coast, and in 1648 the Cossack Dezhnev had (unwittingly) sailed through the strait between Asia and America. Incorporation in the south, however, took longer, for the Russians clashed in lingering conflict with the Chinese and their Buryat vassals, while Cossack penetration into Amuria was for long thwarted by the still powerful China.

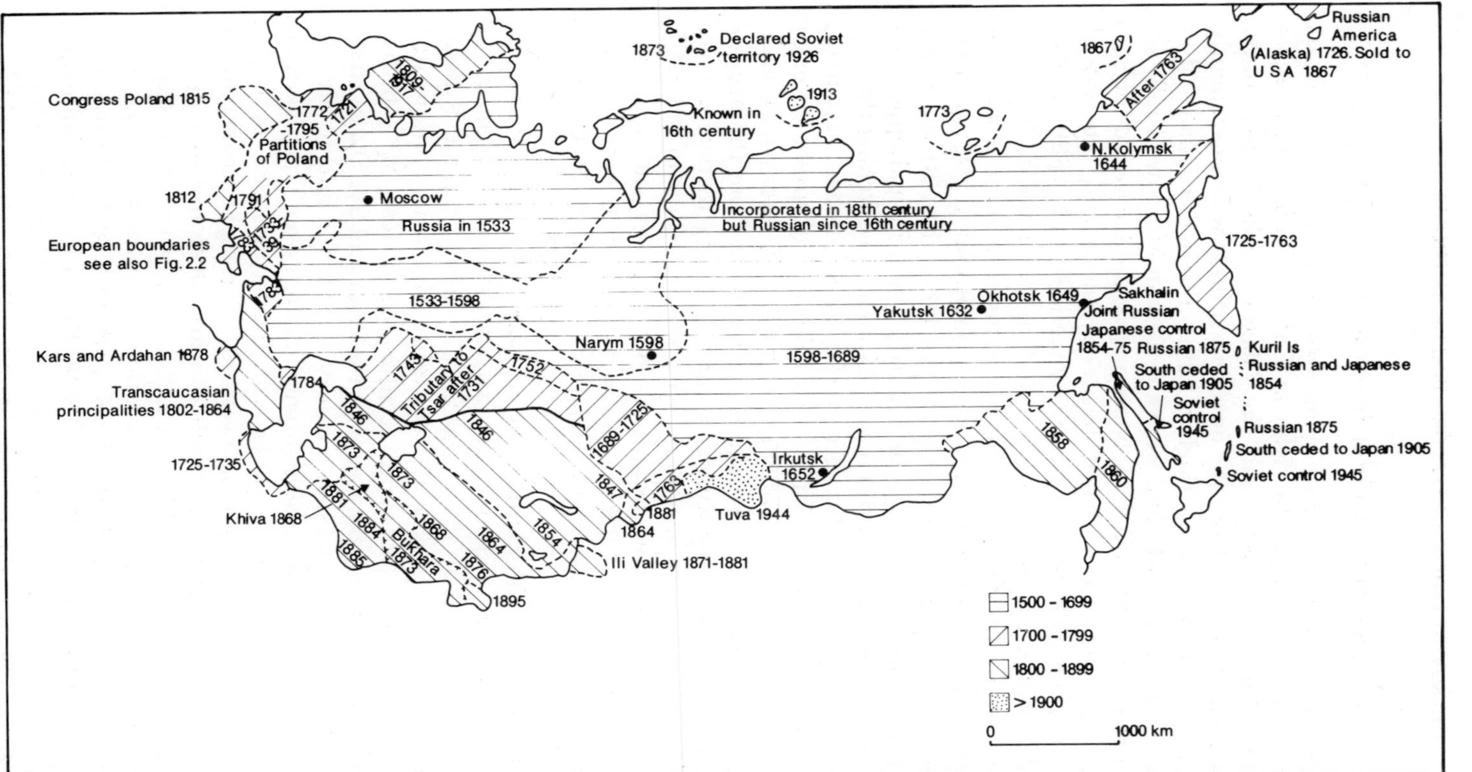

Russian expansion into Siberia was rapid from the late sixteenth century, but incorporation of Central Asia and the Amur lands had to wait until the nineteenth century.

Sources: various Soviet and other historical atlases.

Figure 2.1 *The growth of Russia*

2.2 Territorial growth from the eighteenth century

Although vast territories were being won in Siberia, through its struggles against the powerful Roman Catholic Polish–Lithuanian Commonwealth in the west and the Islamic Turks in the south, Russia remained isolated from the outside world. It was consequently an important occasion when English sailors reached Russian soil at Kholmogory (Arkhangelsk) and were able to visit Moscow (1553–4), to find a strangely self-sufficient, conservative and backward society. The gulf that steadily widened between Russia and the rest of Europe was eventually bridged by the dynamic Peter I (1689–1725), rightly designated *the Great*, who turned a still medieval Muscovy permeated by Asiatic traits towards modernisation and Europeanisation. Peter made, however, only small territorial gains, but long campaigns had taxed Russia's adversaries more than Russia itself, and his successors were to reap the rewards. The principal opponents were all past their prime, with Poland–Lithuania weakening through social decay and Sweden through overtaxing its modest resources, while even the Ottoman Turks were slowly collapsing through excessive conservatism and economic stagnation. In three partitions – in 1772, 1793 and 1795 – Poland was eliminated from the map of Europe, with big gains by Russia, including a long Baltic frontage. Gains on the Baltic were also made from Sweden, while in the south large territories were won around the Sea of Azov and along the Black Sea coast. In 1812 Bessarabia was taken; in 1809 Sweden had ceded Finland; and Russia consolidated its position in Poland in 1815.

During the eighteenth century the Russian position in Siberia was further strengthened, and once the last strength of the remains of Mongol–Tatar power in the steppe had been sapped, the Russian expansion into Central Asia rolled forwards. The Tsars consolidated their position in the nineteenth century as a fear of a British incursion from India into the basins of the Amu and Syr Darya rivers grew. From northern Caucasian territory won in the mid to the late eighteenth century, Russia also tightened its hold over Transcaucasia, in many instances through treaties with local rulers who preferred Russians to Persians or Turks as overlords. The rapid weakening of China during the middle nineteenth century, demonstrated by the major concessions won by Western European powers on its eastern seaboard, encouraged renewed exploration in southern Siberia and a spread into the Amur and Ussuri basins, where large territories were incorporated in 1858–60. By the 1890s Russia had wrested major concessions from China in Manchuria and the Liaotung Peninsula, so that Harbin grew into the largest Russian city in Asia by 1905. A major turn came when Japan, worried by Russian intentions in the Yalu basin and possibly Korea, attacked and defeated the tsar's forces, pushing back the thrusting Russians.

Some historians have suggested that Russian expansion was really a search for warm-water ports, but others have claimed that this was not the primary aim and the real search was concerned with the desire for strategic boundaries. This was sought in particular as a defence against the expansion from the Asian periphery of other European imperial powers, while Russia itself hoped to establish a position from which it might also dominate the peripheral countries of Asia. By the end of the nineteenth century the two main European *imperia* in Asia, Russia and Britain, had established a mutually acceptable system of spheres of interest, notably in Persia and in China, though the two imperial systems avoided direct confrontation, since the Afghan Wakhan strip was agreed as a slender but adequate barrier between them. Much of the Russian spread into northern Asia and even into North America may be seen as territory falling to it by default: remote, inaccessible and at the time not outstandingly attractive in economic terms, Russia simply took what others could not or would not claim. Where real international competition existed, Russia did not do outstandingly well.

2.3 Territorial problems after the Revolution of 1917

The Revolution of 1917 greatly weakened Russia, and several peripheral territories, with or without outside help, sought their independence. Notably important were Finland and the three Baltic republics, Estonia, Latvia and Lithuania. As an outcome of conflict with the newly re-established Polish republic, the latter made gains in western Byelorussia and the western Ukraine (for which the term *Ruthenia* was coined), though not enough to satisfy the truly land-hungry Poles. Rumania seized Bessarabia, the western part of the present Moldavian republic, while Turkey reincorporated Ardahan and Kars, really Armenian districts. Shortlived attempts at independence were made in Byelorussia, the Ukraine and in Transcaucasia as well as by the Ingrian Finns near Leningrad. Japan also strengthened its position in the Far East at Russian expense.

With a tight *cordon sanitaire* drawn round the new Soviet state, interest shifted to within its own boundaries, concentrated on asserting its position within the Arctic basin, where international interest had grown considerably. The Soviets felt the need to declare their ownership of the remote islands off the Siberian coast, claiming the strategically important Franz Josef Land and North Land (discovered only in 1913) as well as tightening their grip on Wrangel

Island and the New Siberian Islands. At the same time, an extensive research and development programme was instigated along the so-called Northern Sea Route.

A further stage of territorial expansion began in 1939. With the German invasion of Poland, the Soviet Union found the opportunity to retake eastern Polish districts roughly to the line suggested as a Polish–Russian frontier by Lord Curzon in 1918, but this time both Lvov and Bialystok were included. In 1940 the three Baltic republics (with German connivance) were annexed and Rumania was forced to surrender Bukovina and Bessarabia, while after a winter war Finland had to concede strategic territory around Lake Ladoga and in Karelia as well as in the far north near Pechenga (Petsamo). The Soviet Union also extracted the right to establish a naval base in the Hanko Peninsula, a commanding position in the eastern Baltic.

The German invasion illustrated only too well the strategic and logistic problems of attacking Russia, with its vast distances and extreme conditions. The German armies, despite their high technology, were not adequately equipped for movement over the poor roads and railways in the western parts of the Soviet Union, while their troops were inappropriately clad to survive in fighting trim through the winter, which in 1941 came early and was severe. The battles confirmed the long-held view of soldiers that if Russian armies are to be decisively defeated, the key battles must be fought and won before Smolensk, the true gateway to Moscow. This decisive defeat was not achieved and the Red Army fell back, luring the Germans into the rapidly expanding dimensions of the interior with its ever-lengthening fronts and vulnerable lines of supply. The Red Army's retreat to more easily defendable interior positions and towards its major war factories beyond the Volga eased its own supply and transport difficulties as it worsened those of the Germans. Trained and equipped to fight and survive through the worst of winter, the Red Army gained strength as the Germans weakened. Trying to maintain a combat presence under conditions for which they were neither trained nor equipped, the Germans ran out of impetus, failing to take three key positions which strategically or politically would have given them the trump card – Leningrad, Moscow and Stalingrad (Volgograd). Overstretched by a war on two fronts and with the initiative lost to them, the Germans were forced back at an ever-growing pace: by mid-1944, the Red Army had advanced beyond the Soviet borders.

2.4 Territorial gains after the Second World War

The defeat of Germany in 1945 returned the Soviet Union not only

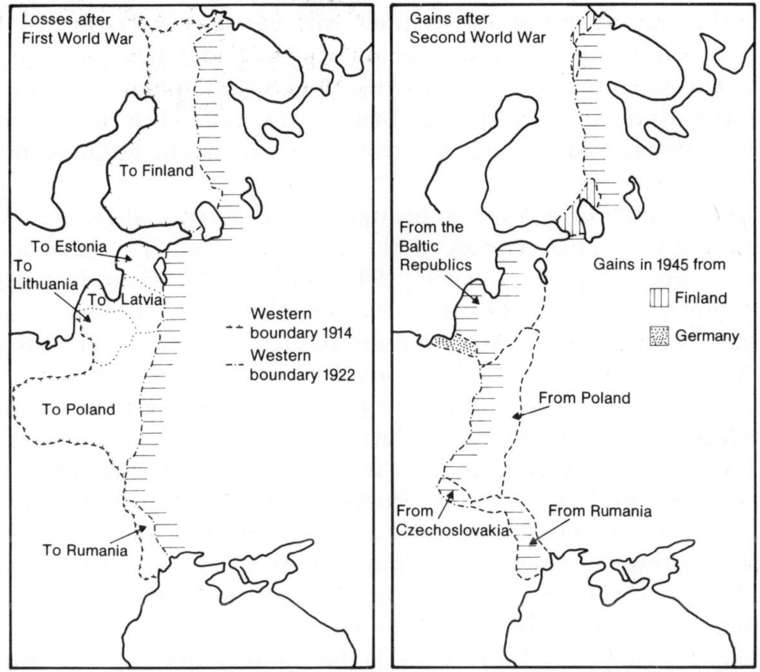

Territorial change this century has been striking – a retreat after the First World War was countered by a new advance after 1945

Figure 2.2 *The western boundaries of the Soviet Union*

to territory gained up to 1941 but also to some important further strategic additions (Figure 2.2). Two key positions were secured: on the north, incorporation of former German East Prussia added the strategic territory around Königsberg (Kaliningrad) and the naval port of Pillau (Baltiysk), from which all three main arms of the Baltic could be commanded; and on the south, the boundary was pushed across the Carpathians, giving Russia command of the major passes from the Ukraine into the Pannonian Plains as well as a powerful strategic hold on the southern Carpathian footslope. The Soviet hold was further consolidated by the incorporation of the port of Ismail and the northern side of the Danube delta. The Red Army advanced deep into Europe and established garrisons and an administration in the Soviet occupation zone in Germany, while Communist regimes in the Eastern European countries created states modelled on the Soviet pattern of society and economy. Stalin hoped that at least some of these states would opt to become members of the USSR, and at the height of his xenophobia all were closely encapsulated into the Soviet political sphere. Their economic potential was as important as their role as a buffer eradicating the springboard from which attacks on Russia had taken place in the past. The westernmost limits of this new Soviet sphere of influence gave the Red Army a strong forward position close to the main Western defence line along the Rhine and only some 50km from the Elbe estuary.

The Second World War also brought Soviet territorial gains in Asia. Straddling the upper reaches of the Yenisey, Tannu Tuva had been a Russian protectorate between 1911 and 1921, when it became independent, but in 1944 it (reputedly) voluntarily joined the Soviet Union as the Tuva Autonomous Oblast. After 1911 Russian influence had grown in Outer Mongolia, which in 1921 had become a Soviet-style republic, and as the Japanese threat to Baykalia from Manchukuo (Manchuria) grew during the 1930s, Soviet interest in Mongolia had developed more strongly. In 1945 China renounced any further interest in the territory, which was then drawn closely into the Soviet orbit, though not *de jure* becoming a member of the USSR. During the waning of Chinese political power in the 1930s, the Soviet Union established considerable influence in Sinkiang (Chinese Turkestan), but in 1941, with the USSR weakened by the German invasion, the Chinese took the opportunity to force out all Soviet influences. Having attacked the Japanese in Manchuria in 1945, the Soviet forces stayed on and held the Liaotung Peninsula until this concession from the Chinese was relinquished in 1955. The Soviet Union also took back all Sakhalin and the Kuril Islands from the defeated Japanese, turning the Sea of Okhotsk into a Soviet lake. As Sino–Soviet relations have deteriorated, Chinese propaganda has made claims to Soviet territory once in Chinese hands, notably in the Ili valley and what Peking calls 'the Great Northwest' and such claims to territory in the Amur valley flared into open conflict in the early 1970s.

2.5 Internal territorial problems

Moscow, the core of Russian nationhood, dominates the Soviet Union as the focus of what is now claimed to be a multinational union of republics. From being the 'Third Rome', focus of the Orthodox Church, it has become the Mecca of Marxist-Leninists and commands the heartland settlement areas of the still expansionist Eastern Slavs. European Russia, overwhelmingly Slav, remains the dominant economic and demographic focus, with two-thirds of Soviet population on one-sixth of the area where over half the total national investment takes place. The first growth under Muscovite leadership drew together the Russian principalities in their own defence, searching for boundaries that could be defended against their enemies, and once these weakened the frontier of settlement, ever moving outwards, became the major element in the territorial growth of the Russian tsardom. There were times and places of remarkably rapid spread – mostly reflected in the occupancy of territory by default, either because there were few or no indigenous people to oppose the Russian penetration or because no other more powerful group could reach it or covet it. Even today, the internal frontier of settlement remains an important element in Soviet life, now translated into essentially economic terms, though along the southern border it has lingering political-geographical undertones.

The modern Soviet state encompasses many different national groups apart from those comprising the original nation-building Eastern Slavs. The rigorous centralism and the way in which the major regions are tied together in economic and political terms suggest the need for a powerful centripetalism to offset a possibly strong centrifugal and divisive force arising from remoteness and from the potential separatism among the less-committed national groups. Much of the difficulty of holding together the country is knowing what is going on. As already noted, communication in everyday affairs across several hours time distance is in itself a problem with centrifugal undertones, to which the remarkable standardisation and uniformity of Soviet life across the country must be seen as a counterpoise. Perhaps the modern rigorously defined 'party line' that has replaced the former demand for unswerving loyalty to the tsar is a way to insure that provincial administrators far from Moscow act as is expected of them.

In some ways we may see the essential political-geographical problem of the Soviet Union as the 'embarrassment of space' and the need to fill it and to use it. After all, in an area ninety times the size of the United Kingdom there are only five times as many people. The reality of this problem is immediately clear when we think that 78 per cent of the area has less than ten persons per km^2 – the average density in this belt is, in fact, a mere two persons per km^2, virtually meaning it is uninhabited, whereas in contrast little more than 5 per cent of the total area has more than fifty persons per km^2. How does a state supply the infrastructure of modern life in this sort of situation? Over 70 per cent of the total population lives in European Russia, the homeland of the Eastern Slavs, that comprise together over three-quarters of the total Soviet population. The next largest major group, the Turkic peoples, form only a little over 15 per cent of the total. Of 104 'nationalities' only twenty-three numbered more than one million people, so clearly the great proportion are numerically extremely small and often scattered over vast areas, while many are also at a low level of economic and cultural achievement. Most important is the fact, however, that some of the numerically largest and most advanced of the non-Slav peoples live on the periphery of the Soviet Union and have close relatives beyond the Soviet borders, thus creating further political-geographical complications. It is perhaps interesting that the Russian spread across Siberia was hindered more by physical obstacles than by resistance from indigenous peoples, unlike the way the Indians opposed European penetration along the American frontier. In southern Siberia and in Amuria it was under a transient Chinese pressure that native peoples briefly but irresolutely opposed the incoming Russians. Siberia was conquered more through the vodka bottle than through the six-shooter! In Central Asia the story was different: from the outset the steppe nomads put up a fierce resistance, and even the decaying Islamic khanates in the south waged a rugged struggle against the relentless pressure of the Cossack armies. A varied story reflects the takeover of Transcaucasia, where the Russians were often preferred as overlords to Turks and Persians, though many mountain peoples, fiercely independent, fought long against the tsar's armies. Resistance varied on the western limits of Russia in Europe, with parties often changing sides, though the Poles have formed the most concerted opposition to a westward spread of Russian power. The numerically small Baltic peoples successfully sought outside help against Russian domination for a brief interlude between the world wars. It is also noteworthy that not even the Eastern Slavs were always united – shortlived attempts at independence have been made by both the Ukrainians and the Byelorussians.

2.6 The territorial-administrative organisation (see Figure 2.3)

As an aspect of internal political geography, we must also consider how the territorial-administrative structure of the country has developed. Old Russia was governed from the remains of the principalities absorbed by the spread of Muscovy, but in 1708 Peter I replaced these by large *governments*, whose number increased as more territory was added. Once formed, however, the reform of their boundaries, seldom undertaken, failed to keep pace with the broadening economic and social horizons, so that by 1917 the Tsarist Empire was divided into some seventy-eight *governments* and eighteen *provinces*, badly in need of reform, the sheer magnitude of which had no doubt deterred successive tsars.

The 1917 Revolution created the Union of Soviet Socialist Republics, an apparently elaborate federal structure but in reality remaining almost as centralised as the tsardom that had been swept

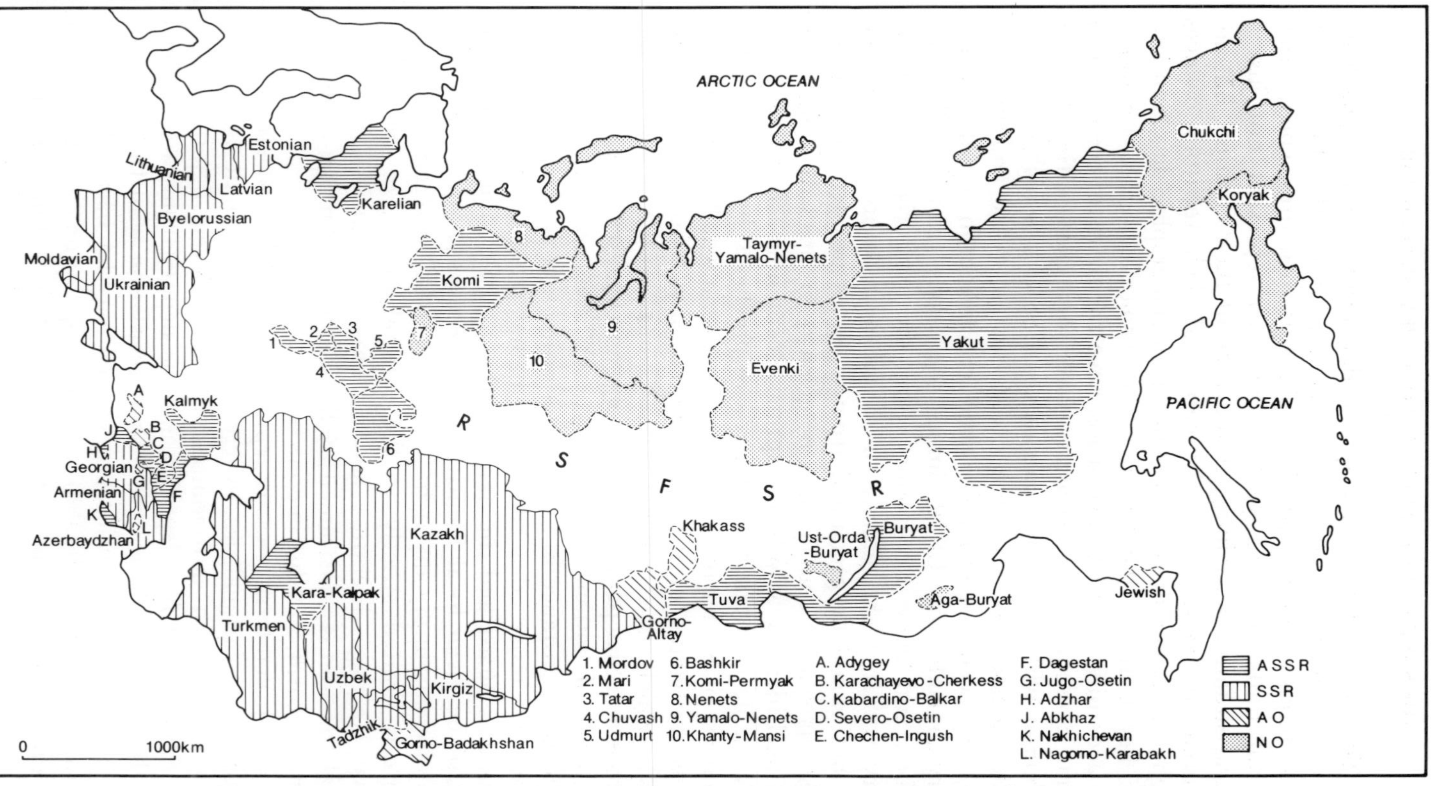

The complex territorial-administrative map arises from the several levels of ethnic autonomy.

Source: territorial-administrative map supplied by the Soviet Embassy.

Figure 2.3 *Administrative-territorial structure*

away. The importance of the spatial dimension was recognised by being woven into the core of the administrative structure of the Union in terms of both political and planning structures. The internal territorial units of local government now represent not just a historical situation pertaining at the time of their creation, but are monitored and modified as the nature of their territory changes through the dynamics of economic and demographic development.

The key units of the Union are the so-called union republics, which according to the Constitution have the right to secede if they wish and consequently must always have some part of their borders in common with a country outside the USSR. It is highly unlikely that the question of secession could ever be raised and even less likely to succeed if it were. The need for a border in common with a country outside the USSR limits the likely number of union republics, and there are groups within the territorial heart of the country which might already or in the near future warrant such status but would be precluded from it by virtue of their location. Such status can, of course, only be given to groups with a high level of national identity as well as impressive economic achievement. For groups unable to attain these criteria or precluded by location deep in the heart of Soviet territory where secession would be impossible, there is the status of the Autonomous Soviet Socialist Republic (ASSR), which is the highest level some of the advanced and numerous non-Slav peoples of the Volga and Ural will ever reach. For lower levels of national attainment nevertheless warranting recognition (in some instance in Siberia as a protection against being overwhelmed by more advanced and dynamic groups), there are the autonomous *oblast* and the national *okrug*, while some small groups may also be given local identity where necessary. Such status does not in any way imply any right to self-rule or self-determination: the concept is limited strictly to the cultural sphere, to the use of national language, literature and customs. Nevertheless, even this cultural freedom is constrained by Soviet ideology in that it must be 'national in form, socialist in content'.

These national territories are represented in central government by the Soviet of Nationalities, where each union republic has thirty-two deputies, each autonomous republic eleven deputies, each autonomous *oblast* five deputies and one deputy from each national *okrug*. It has an equal right to initiate legislation with the Soviet of the Union, whose deputies are elected on the basis of one for approximately each 300,000 of the population irrespective of nationality. Legislation is adopted if passed by a simple majority in each chamber. These two chambers meet, however, only occasionally each year to ratify the work of government done through the powerful Presidium of the Supreme Soviet to which they belong, but run in effect by the Politburo.

The boundaries of the various types of national territories are drawn to reflect the limits of the ethnic groups they represent and may be changed only with the group's consent and in response to changes in the pattern of ethnic distribution. Other territorial-administrative units can be changed generally more readily. Although Lenin set the precedent that the boundaries of the republics should be regarded as immutable when he rejected the proposals of the 1921 Kalinin Commission on electrification, some changes have taken place. After incorporation of the Baltic republics in 1940, areas with Russian majorities on the borders of Estonia and Latvia were transferred to the Russian Soviet Federated Socialist Republic (RSFSR), while Lithuania received its majority area around Vilnyus. In 1945, when Bessarabia was taken from Rumania, the Moldavian SSR was enlarged, but Ukrainian ethnic areas previously attached to it to make it viable were returned to the Ukrainian SSR. Likewise the physical and economic unity of the Fergana Valley was broken to satisfy national aspirations, though some industrial trusts have been administered on a multi-republic basis. An artificially created and unsuccessful national area is the Jewish autonomous *oblast* near Birobidzhan in the Far East, designated in the early 1930s as a 'national home' for Soviet Jews, who still form less than 10 per cent of its population. During the Second World War some national areas were annulled because their inhabitants were accused of collaboration with the Germans, notably the Crimean Tatar ASSR and the Volga German ASSR. Some Caucasian peoples and the Buddhist Kalmyks suffered the same fate, but their territories were reinstated in 1957. Although progress has usually been up the ladder of autonomy, Russian immigration coupled to changing political conditions brought the reduction in the status of the Karelian Finns from a union republic (SSR) to an autonomous republic (ASSR) in 1956. Most national territories do, of course, contain large minorities of other groups, particularly Russians, Ukrainians or adjacent peoples. In some cases these are more numerous than the titular group, as, for example, in the Kazakh SSR, where the Kazakhs themselves form well under half the total.

Areas of the larger union republics not occupied by lower-ranking national districts are divided into territorial units that serve both administration and economic planning needs. These are based on structural concepts formulated in the 1920s on which economic development programmes might be pinned. The basic unit is the *oblast* or, in the less-developed parts of Siberia where much larger units of this type are found, the *kray*. Each is subdivided into *rayony*. Each *oblast* or *kray* is seen as the hinterland of its main industrial town (usually a route focus) and 'proletarian centre', serving its needs as far as possible, and the area may range from 5,000km^2 to over 1 million km^2. In each district the aim of the planners is to create a diverse economy, though (perhaps paradoxically) local specialisation should not be overlooked, but it must not come to dominate the local economy. The aim is also to make each district as self-supporting as possible to try to eliminate wasteful transport, particularly cross-hauls, with special attention given to ensuring local energy supplies. Achievement of completely independent regional economies is, of course, impossible because of the geographical inequality in the distribution of natural resources, which consequently tend to select parts of the country for particular development (e.g. cotton-growing in Central Asia). This emphasis is also explainable in terms of national defence through establishing regional autarky and levelling out industrialisation between the developed and less-developed parts of the country. There is also reputedly economy in unnecessary transport and the need to provide additional routes, savings which may be diverted to more 'productive' operations. Nevertheless, the local district developments have, say Soviet planners, to be seen in an 'all-union' perspective, showing that they make a real contribution to the over-all national welfare. It is striking how the number of *oblasti* has increased over time as they have been subdivided to serve the growing intensity of economic development and the numerous changes in their boundaries reflect the intentional flexibility to adapt to spatial change in the economic geography of the country.

2.7 Economic regionalisation

The needs of medium-scale economic management can well be built into the pattern of *oblasti* along with the local government functions, but for broader-scale and longer-term economic planning more able to mirror national change larger units, suitable for statistical as well as planning purposes, have been designed. While basically the union republics might form such a system, their widely disparate characteristics of area, population and economic potential do not make them ideal. Consequently an economic regionalisation without any administrative government function has been delineated, though the pattern has been varied from time to time.

The idea of large fiscal and statistical units for planning goes back to 1921 when the Kalinin Plan for major generating divisions for the electrification of Russia was published. These regions were based on each having a major centrally located source of electricity generation, but they were also proposed as a system of general economic planning regions. Although drawn only crudely with straight-line boundaries, each region was designated by its main economic criterion – 'predominantly industrial', 'predominantly agricultural' or 'forest region'. There was much to commend in the scheme – perhaps more so than in some later proposals – but it was rejected because it did not recognise sufficiently the integrity of the boundaries of the union republics. For example, the Ukrainian and Russian sectors of the Donbass coalfield were combined in a large southern region, while the then small Byelorussian republic was put into the western region of the RSFSR. Such contraventions of Lenin's 'national principle' were inadmissible because, for example, Byelorussian national development would be overshadowed by its larger neighbour, as equally would be the proposed amalgam between the embryonic Transcaucasian republics and northern Caucasia, since the latter was also part of the giant RSFSR. Nevertheless, during the 1920s two prototypes of the new concept were set up in the industrial Ural and in agricultural northern Caucasia, while the administrative-territorial pattern that appeared with the New Economic Policy bore its imprint. Until the 1930s, however, there was a clear correspondence between the economic planning regions and the majority of administrative units' boundaries, but as the number of the latter increased and their size decreased, the need for special large regions as a structure for long-term planning became more pressing. While the smaller units were able to give the guidance expected of them to agriculture and to small-scale industry, planning of large-scale industry centrally directed by ministries became more difficult as their number increased.

The outcome was the development of major economic regions, the so-called *makrorayony*, within whose boundaries a number of administrative districts were drawn together (Figure 2.4). In a way this concept also fitted the *gigantomania* that dominated Soviet thinking between the two world wars, though the huge plants established in this phase for prestige reasons did little to ease the pressure on the transport system. Each large *makrorayon* was expected to develop a high level of self-sufficiency, not only to reduce the burden on transport, but also to form part of a system of 'bulkheads' in case of war, principles already employed in the design of the territorial-administrative units. Soviet commentators criticised them as never likely to be able to reach the expected levels of self-sufficiency because in their design 'geographical realities' were too often ignored, but then no single *makrorayon* could be expected to contain all the 1,500 or more substances needed to feed a self-sufficient modern industry, for natural resources are too unequally distributed. Never-

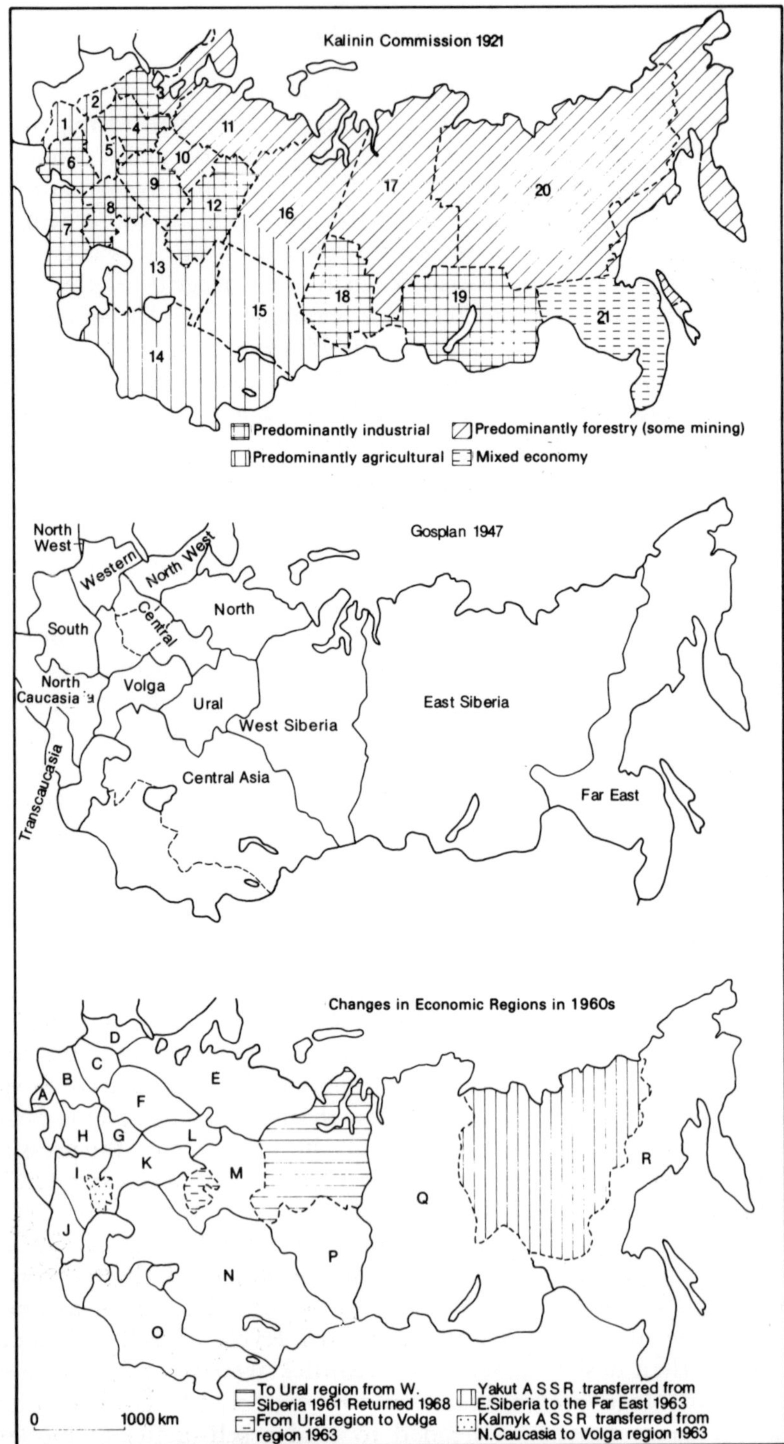

The attempts to define economic regions have produced many different patterns, though a more or less conventional usage is now accepted based on the Gosplan regions of 1947–68.

Figure 2.4 *Boundaries of planning regions*

Figure 2.4 *contin.*

Sources: various Soviet maps.

theless, a number of major industrial projects did ultimately owe their origin to the *makrorayon* concept and the idea of regional autarky. Some projects were part of a desire to disperse essential strategic industries while levelling out regional disparities in development, as (for example) the steel plants at Bekabad in Central Asia and Komomsolsk in the Far East, and strategic undertones were certainly vital. The difficulties are reflected by the Ural region, which in so many ways came close to the ideal and yet was deficient in vital coking coal for its heavy industry. Siberian regions, though immensely rich for industrial development, could never hope to produce sufficient foodstuffs for their own needs. On the other hand, Central Asian republics were needed to produce crops not otherwise possible in the Soviet Union and consequently had to import food from elsewhere. Some students have suggested self-sufficiency was not equally pursued in all regions for political reasons – the Central Asian republics were, so it is claimed, purposefully made dependent by the planners on considerable *outside* food supplies, because their political reliability and loyalty to the Soviet principle were under suspicion.

After the war, in 1947, the *makrorayon* appeared again when the country was divided into thirteen major regions that included newly incorporated territories. The division came, however, under considerable criticism, with a common complaint from Soviet observers that thirteen regions were too few to mirror the true diversity of the economic geography of the country and that in some instances clear affinities of territory to particular centres appeared to have been ignored in drawing the boundaries. The number of *makrorayony* was later increased to fifteen, eventually to seventeen and after 1963 to nineteen, though some studies had recommended more than twenty such units. The changes tended to reflect a shift towards an acceptance of more regional specialisation, for example the splitting of the central industrial region from the strongly agricultural central black-earth region or by dividing the Ukraine into the industrial Donets–Dnepr region and the mostly agricultural south and south-west. Boundaries were also changed through the 1950s as the reorganisation of the *oblasti* was undertaken, while a decade later major shifts were made in the boundaries of the *makrorayony* themselves. Whatever boundaries were selected and possibly whatever number of *makrorayony* defined, the criticism that they concealed areas with a distinct character and many interregional differences would doubtless still have been at least partly true. Nevertheless, they became the most commonly used divisions for all kinds of regional studies and descriptions in Soviet literature. It is perhaps a tribute to their design that however their boundaries have shifted, their core areas have remained fixed and readily definable regional foci, the central pegs around which the regional framework of the Soviet Union has been built.

Under Stalin the large *makrorayony*, despite heated debate about them, played little part other than for statistical and planning purposes, since the economy was highly centralised and managed in vertical sectors, so that there was effectively little play for the horizontal management that might have been expressed through them. The Khrushchev era decentralised the economic system, with the industrial ministries largely disbanded and control handed to around a hundred regional economic-administrative commissions (*sovnarkhozy*). These usually comprised a single *oblast*, *kray*, ASSR, or a small union republic, but in some instances the combination occurred of three or four such units into a single *sovnarkhoz*. Each *sovnarkhoz* possessed full planning, managerial and budgetary responsibility for all aspects of economic activity within its territory; only major construction projects of national interest and strategic industries remained under central control. Local loyalties, however, lobbied for boundary changes in the *sovnarkhozy*, but the total number fluctuated only relatively slightly. Nevertheless, they were widely disparate in form: the Kiev *sovnarkhoz*, for example, had 8 million people, Kamchatka 200,000; the vast Yakut ASSR, almost the size of peninsular Europe, formed one *sovnarkhoz* and so did the small Karbadino-Balkar ASSR. Economic structure ranged from heavy industry in the Donetsk *sovnarkhoz* to the then forest economy of the Tyumen *sovnarkhoz*. Certainly the system gave a new emphasis to regional specialisation, but the retention along with it of the large *makrorayony* suggested that the older principle of integrated but diversified regional development had not really been absolutely abandoned.

The *sovnarkhoz* system quickly showed excessive fragmentation of economic control, with local loyalties and needs often getting too great a priority over the 'all-union' perspective. Feeling moved towards larger units, since the largest *sovnarkhozy* had showed the best performance, and by 1963 the 105 *sovnarkhozy* had been consolidated into forty-seven industrial management regions, with the *makrorayony* modified to take up the new arrangement. Most significantly, however, the creation of state production committees responsible for individual branches of industry on a national basis marked a return towards centralisation.

The changes had produced an elaborate hierarchy for the management and control of the economy, with a good deal of overlap. Of great significance was the view now expressed that the boundaries of the union republics were losing their former importance as technological and economic advance demanded closer interrelationships and co-operation between them. A new interest was generated in so-called 'territorial production complexes', which were to be 'historically formed and objectively existing', with many schemes for their identification and delineation put forward well into the 1970s. Since 1965, however, the preference has swung to a centralised and vertical direction of the economy through some twenty-eight ministries, so that all economic regional councils except those of the union republics have been swept away. It would be interesting to know how far this change of heart in the Politburo and in Gosplan has been conditioned by application of computer techniques and the building of an effective computer network, which would unquestionably make central direction more feasible than ever before. On the other hand, the shift back to central, vertical direction may reflect a growing mistrust that decentralisation was leading to a weakening of the links between Moscow and some of the national groups or even distant parts of the RSFSR itself.

From this complex story of change emerges the difficulty and dilemma of effectively controlling the vast Soviet territory: should this be achieved by a detailed approach through myriads of small units that hide regional differentiation? Much of the difficulty becomes confused) or by relatively few but consequently vast units that hide regional differentiation. Much of the difficulty arises probably from the disparate character and level of development of the various regions, but there has also been debate about what sort of regions are needed, based largely on ideological arguments. The choice has been between the relative merits of homogeneous units (the usually accepted criterion in the West) and the heterogeneous units, conditioned by strategic and logistic arguments, particularly the reduction in transport effort, apparently most favoured in the Soviet administration. Perhaps one of the most striking anomalies in the definition of territorial units has been consistent disregard of the special problems and nature of the Soviet Arctic. Handling the special circumstances of the northlands has been left to organisations like the Administration of the Northern Sea Route (*Glavsevmorput*) or the Far Eastern Construction Trust (the sinister *Dalstroy*) and to a special ministerial responsibility for major construction projects set up late in 1979. The northlands are also identified especially for work and pay conditions. Despite frequent criticism from within the Soviet Union of this situation, the reason seems to underlie the idea of heterogeneity and the need for a strong 'proletarian centre' in each territorial unit, which in the case of Siberia can only be found along the southern railway zone.

2.8 Where to follow up this chapter

Many good historical surveys of Russia exist – among these are Clarkson, L. D., *A History of Russia* (Longman, London, 1962), Florinsky, M. T., *Russia – A History and an Interpretation*, 2 vols (Macmillan, New York, 1953), Kochan, L., *The Making of Modern Russia* (Cape, London, 1962), Parker, W. H., *An Historical Geography of Russia* (University of London Press, 1968), Riasanovsky, N. V., *A History of Russia* (Oxford University Press, 1963), and Sumner, B. H., *A Survey of Russian History* (Duckworth, London, 1944). Also useful are Kerner, R. J., *The Urge to the Sea: the Course of Russian History* (University of California Press, Berkeley, 1946), and the four volumes by Vernadsky, G., on the early history of Russia. Other works include Portal, R., *The Slavs – A Cultural and Historical Survey of the Slavonic Peoples* (Weidenfeld & Nicolson, London, 1969), and Wesson, R. G., *The Russian Dilemma – A Political and Geographical View* (Clark, New Brunswick, 1974). The exploration of Russia is covered in Berg, L. S., *Die Geschichte der russischen geographischen Entdeckungen* (Haack, Leipzig, 1954).

The internal organisation of territorial-administrative units is covered by Dewdney, J. C., *The USSR*, Studies in Industrial Geography (Hutchinson, London, 1978), and his *Patterns and Problems of Regionalisation in the USSR* (University of Durham Research Papers 8, 1967). On the territorial production complex, see Kozlov, I. V. (ed.), *Novyye territorialnyye Kompleksy SSSR* (Moscow, 1977), and Lonsdale, R. E., 'The Soviet Concept of the Territorial-Production Complex', *Slavic Review*, 24, 1965, pp.466–78. A most recent work is Pallot, J. and Shaw, D. J. B., *Planning in the Soviet Union* (Croom Helm, London, 1981).

3

The Soviet People

In population terms the Soviet Union, with 265 million people (1980), ranks third in the world after the Chinese People's Republic (835 million) and India (598 million). In relation to its area, however, the population is relatively modest: although it is twice the area of the USA, its population is little more than a fifth larger, and despite being ninety times the area of the United Kingdom, the population is not quite five times greater.

During the nineteenth century the population of Russia trebled from about 45 million in 1800, overwhelmingly the result of natural increase. Although there have been marked fluctuations in the rate of increase, since the tsarist census of 1897 the population of the present area of the Soviet Union has virtually doubled. Before the First World War, by contemporary Western European standards, there was still vigorous growth, with a high birth rate dampened, however, by a high mortality, particularly of infants. During the war, revolution and subsequent civil war, fertility fell substantially and there was even higher mortality, which probably accounted for a 'shortfall' of about 25–30 million people, arising from about 16 million or more additional deaths and a loss of 10–12 million births as well as anything up to 2 million emigrants. When stability returned in the 1920s, there was a rapid return to a high birth rate (partly through official encouragement) and deaths fell to below the 1913 rate, but the end of the decade saw a more marked decline in the birth rate rather than a further fall in the death rate. The trend was thus to a lower rate of natural increase, though less dramatically so than in Western Europe. The Second World War had a drastic effect: estimates of 14–20 million war deaths have been made and a shortfall of between 15–20 million births, with the resultant population

Table 3.1 *Growth of population*

Date	Area within boundaries of time (million km²)	Total population (millions)	Urban (millions)	Rural (millions)
1897	22.3	128.2	20.1 (16%)	108.1 (84%)
1926	21.7	147.0	26.3 (18%)	120.6 (82%)
1939	21.7*	170.6	56.1 (33%)	114.5 (67%)
	22.1†	190.7	60.4 (32%)	130.3 (68%)
1959	22.4	208.8	100.0 (48%)	108.8 (52%)
1970	22.4	241.7	136.0 (56%)	105.7 (44%)
1979	22.4	263.4	164.4 (62%)	99.0 (38%)
1980	22.4	265.5	167.3 (63%)	98.2 (37%)

* Boundaries of 1926.
† Boundaries including annexations.
Source: *Narodnoye Khozyaystvo SSSR* (Moscow, various years).

in 1946 probably anything of the order of 30–40 million less than it would otherwise have been; in 1950 Soviet population was still 15.6 million below the 1940 level, which was not re-attained until 1955.

After the Second World War population growth swung upwards and the rate of natural increase rose until the early 1960s, since when it has fallen back, and by the mid-1970s appears to have flattened out. Partly this has been the effect of the 'thin generations' of the war years coming up to parenthood. The birth rate fell fairly consistently until about 1970, and has since fluctuated with a slight upward trend, but the death rate has also fallen consistently until the late 1960s, since when there has been an upward trend as the ageing of the population has become more pronounced. Over all, natural increase has consequently tended to decline, having reached 4 million annually in 1960, but subsequently falling back to a little over 2 million, suggesting that an annual growth of less than 1 per cent in the near future can be expected.

Information, albeit rather limited, on age and sex structure suggests that the events of this century have expectedly left a deep impress in a markedly uneven distribution between the different cohorts. The two world wars and the civil war left a serious distortion in the balance between the sexes in the age groups in which male combat fatalities occurred. In the lower age groups the small classes of the war years and the large ones of the brief 'baby

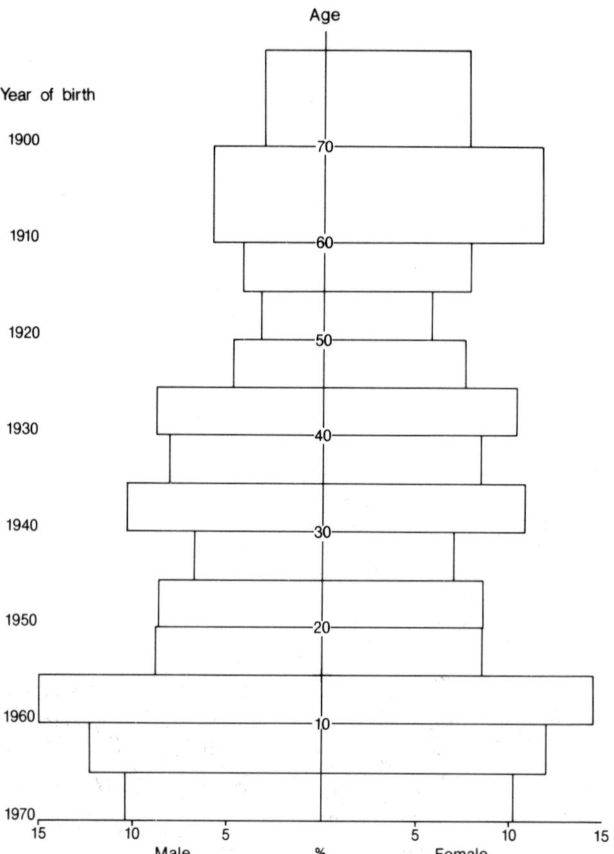

The turmoil of events this century has been reflected in the unbalanced nature of the population pyramid. Notable are the heavy losses through war and civil commotion on the male side, while the falling birth rate after 1960 is also apparent.

Source: *Soviet Statistical Yearbook.*

Figure 3.1 *Age and sex structure of the Soviet Union, 1970*

boom' of the 1950s stand out, while the tapering base in the youngest age groups marks the recent decline in the birth rate. The continuing relatively low birth rate compared with the past has considerable long-term implications for economic development, especially the smaller annual additions to the labour force. Likewise the pyramid suggests larger classes in the near future in the older age groups, with all the implications implicit for the provision of welfare services. The trends seem, however, consistent with experience elsewhere in countries with increasing affluence, industrialisation and rising proportions of urban population. It is perhaps not without

Table 3.2 *Births, deaths and natural increase by republics per 1,000 of the population*

	1940			1978		
	Births	Deaths	Natural increase	Births	Deaths	Natural increase
USSR	31.2	18.0	13.2	18.2	9.7	8.5
1. RSFSR	33.0	20.6	12.4	15.9	10.3	5.6
2. Ukraine	27.3	14.3	13.0	14.7	10.7	4.0
3. Byelorussia	26.8	13.1	13.7	15.9	9.1	6.8
4. Uzbekistan	33.8	13.2	20.6	33.9	6.9	27.0
5. Kazakhstan	40.8	21.4	19.4	24.4	7.4	17.0
6. Georgia	27.4	8.8	18.6	17.7	8.0	9.7
7. Azerbaydzhan	29.4	14.7	14.7	24.9	6.7	18.2
8. Lithuania	23.0	13.0	10.0	15.3	10.0	5.3
9. Moldavia	26.6	16.9	9.7	20.1	9.8	10.3
10. Latvia	19.3	15.7	3.6	13.6	12.4	1.2
11. Kirgizia	33.0	16.3	16.7	30.4	8.1	22.3
12. Tadzikstan	30.6	14.1	16.5	37.5	8.3	29.2
13. Armenia	41.2	13.8	27.4	22.2	5.5	16.7
14. Turkmenistan	36.9	19.5	17.4	34.4	8.0	26.4
15. Estonia	16.1	17.0	–0.9	14.9	12.2	2.7

Source: *Narodnoye Khozyaystvo* (1978 edition).

political meaning that over 80 per cent of the present population has been born since the Revolution.

Over-all national figures for growth and the age and sex structure in the Soviet Union obviously conceal considerable regional variation. From material available, the pattern of natural increase regionally suggests the lowest levels lie in the European centre around Moscow and in the black-earth belt to the south, also characteristic of the north and north-west and the Baltic republics of Latvia and Estonia. Natural increase is a little greater in Lithuania, Byelorussia and the Ukraine and appreciably higher in Moldavia (on the same level as the neighbouring, related Rumanians) (Table 3.2). In the RSFSR east of the Volga, in the Ural and Western Siberia, natural increase is also higher than in central European Russia. In Eastern Siberia and the Far East the level is still higher; a Soviet explanation of this pattern is that it arises from the younger structure of the population in the many economic growth poles of these regions. By far the highest levels of natural increase occur in Transcaucasia and especially in the Central Asian republics, with levels of two or three times the national average: Soviet sources suggest this arises from the considerably different cultural traditions among the non-Slav population of these areas compared with the essentially Slav population of European

Russia. In the long term, of course, this could cause not only a considerable shift in the spatial distribution of population but also alter the long-standing relationships between the different ethnic groups.

Falling birth rates diminish the numbers entering the labour market some fifteen years later and this problem also has its regional variations. Figures suggest the considerable long-term decline in labour availability in the more urbanised and developed parts of European Russia, though this might be offset by migration from Siberia, Central Asia and Transcaucasia, where the annual labour supply is expected to remain more buoyant. In the early 1960s, for example, the annual numbers of new recruits to the labour force fell sharply, with less than three-quarters of the young people in the 15–19-year group compared with the second half of the 1950s. The shortage was overcome by the first major cuts in the armed forces, and it has been suggested that Soviet attempts in the 1970s to reach *détente* in Europe may also be conditioned by similar reasons. At the same time, there has been a serious effort to increase the already high participation in the labour market. Some migration may also be necessary, because employment prospects in small towns are mostly less favourable than in the big towns, while heavy industrial towns offer poor prospects for female employment. On the other hand, improvements in social welfare, notably better pension opportunities for farm workers, have helped shake out less productive members in some overmanned sectors. An abundant labour supply is regarded as the key to a high rate of economic growth, though this may now be less true, particularly if the shift from productive industry to the service sector accelerates and if more sophisticated techniques and equipment are introduced. Even so, it looks as though demand will long outstrip supply in the labour market, with some serious regional deficiencies. In Soviet thinking, recruitment of migrant labour from neighbouring countries is totally unacceptable and nowhere in Comecon, despite labour-supply difficulties in several areas, has foreign labour like the *Gastarbeiter* of West Germany been recruited. Such migrant labour is only employed temporarily on special projects.

3.1 **The ethnic problem** (see Table 3.3)

The reader will already have grasped that not all Soviet citizens are Russians, nor are they all Slavs (Table 3.3). The 1970 census enumerated over 800 ethnic denominations comprising about 122 nationalities, speaking 114 languages divided between over 300 dialects. The final published census list comprised 104 national

Table 3.3 *Ethnic structure of the USSR*

Ethnic group	Major ethnic affinity	No. in millions 1959	1970	1979	% change 1959–70	1970–9
		Peoples with Union Republics				
Russian	Slav	114.11	129.01	137.39	13.1	6.5
Ukrainian	Slav	37.25	40.75	42.35	9.4	3.9
Byelorussian	Slav	7.91	9.05	9.46	14.4	4.5
Latvian	Baltic	1.40	1.43	1.44	2.1	0.6
Lithuanian	Baltic	2.32	2.66	2.85	14.6	6.9
Estonian	Finno-Ugric	0.98	1.00	1.02	2.0	1.3
Moldavian	Rumanian	2.21	2.69	2.97	21.7	10.0
Georgian	Caucasian	2.69	3.24	3.57	20.4	10.0
Armenian	Armenian	2.78	3.55	4.15	27.7	16.6
Azerbaydzhani	Turkic	2.94	4.38	5.48	48.9	25.0
Turkmen	Turkic	1.00	1.52	2.03	52.0	32.9
Uzbek	Turkic	6.01	9.19	12.46	52.9	35.5
Tadzhik	Iranian	1.39	2.13	2.89	53.2	35.7
Kazakh	Turkic	3.62	5.29	6.56	46.1	23.7
Kirgiz	Turkic	0.96	1.45	1.91	51.0	31.3
		Peoples with ASSRs				
Abkhar	Caucasian	0.065	0.083	0.091	27.6	9.6
Bashkir	Turkic	0.98	1.24	1.37	26.5	10.6
Buryat	Mongol	0.25	0.31	0.35	24.0	12.1
Dagestani*	Dagestani	0.94	1.36	1.66	44.6	21.4
Kabardino }	Caucasian	0.20	0.28	0.32	40.0	15.0
Balkar }	Turkic	0.042	0.06	0.066	42.8	10.0
Kalmyk	Mongol	0.10	0.13	0.15	30.0	7.3
Kara-Kalpak	Turkic	0.17	0.23	0.30	35.3	28.4
Karelian	Finno-Ugric	0.16	0.14	0.13	−12.5	−5.5
Komi	Finno-Ugric	0.28	0.32	0.33	14.3	1.5
Mari	Finno-Ugric	0.50	0.59	0.62	18.0	3.8
Mordov	Finno-Ugric	1.28	1.26	1.19	−1.6	−5.6
Osetin	Iranian	0.41	0.48	0.54	17.0	11.1
Tatar	Turkic	4.96	5.93	6.32	19.5	3.5
Tuvinian	Turkic	0.10	0.13	0.17	30.0	19.4
Udmurt	Finno-Ugric	0.62	0.70	0.71	12.9	1.4
Chechen }	Caucasian	0.42	0.62	0.76	47.6	23.3
Inguish }	Caucasian	0.10	0.15	0.19	50.0	17.7
Chuvash	Turkic	1.47	1.69	1.75	14.9	3.4
Yakut	Turkic	0.23	0.29	0.33	26.1	10.8
		Other groups				
Eveni	Tungus-Manchurian	0.009	0.012	0.012	33.3	2.5
Evenki	Tungus-Manchurian	0.025	0.025	0.027	0	9.6
Chukchi	Palaeo-Asiatic	0.012	0.013	0.014	16.6	2.9
Koryak	Palaeo-Asiatic	0.0063	0.0075	0.0079	19.0	5.3
Yukagir	Palaeo-Asiatic	0.0004	0.0006	0.0008	50.0	33.3
Dungan	Tibeto-Chinese	0.022	0.039	0.052	77.2	33.3
Jewish	Semitic	2.26	2.15	1.81	−4.9	−15.8
German	Teutonic	1.62	1.84	1.94	13.6	4.9
Polish	Slav	1.38	1.17	1.15	−15.9	−1.4
Greek	Greek	0.30	0.33	0.34	10.0	2.0

* Dagestan peoples include Avars, Lezgins, Dargintsy, Kumki, Tabasarany, Nogay, Rutultsy and Aguly.
Source: *Narodnoye Khozyaystvo SSSR* (1970 and 1979 editions, Moscow).

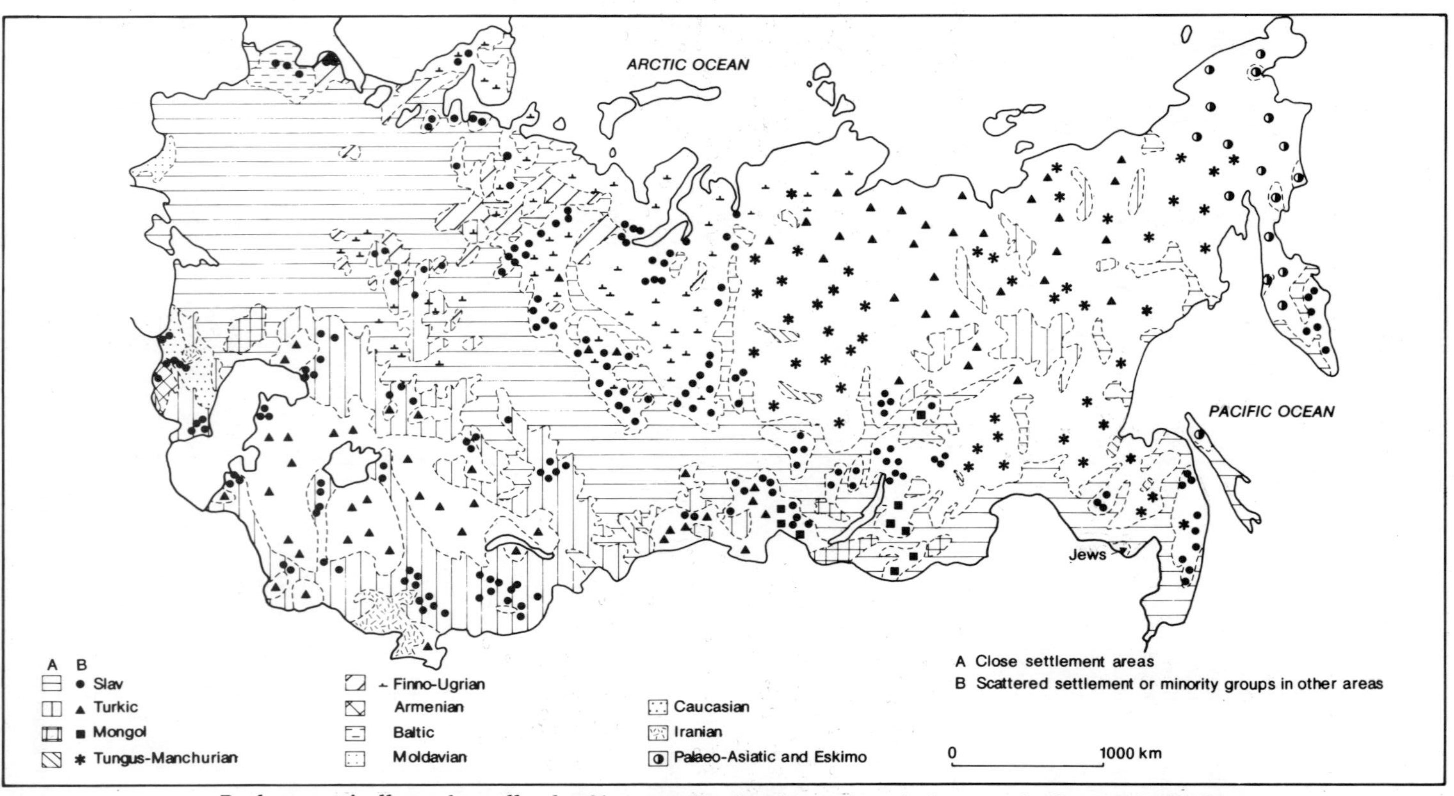

Both numerically and areally the Slav peoples dominate the ethnic scene in the Soviet Union.

Sources: maps from various Soviet atlases.

Figure 3.2 *The ethnic map of the USSR*

groups (simplified to ninety-three groups in the 1979 census), defined on a composite basis which gives language a particularly important role, but also includes culture, history and even religion, though some groups are recognised simply because they have regarded themselves as separate over a long historical period. Physical type plays little role. In tsarist times almost no recognition was given to these different groups and in early enumerations non-Slavs were not even counted. The Revolution made, however, a considerable issue of the 'national question' and the Constitution defined the rights of national groups, quickly followed by definition of the first union republics and other national areas, whose boundaries were given an almost sacrosanct character (see Chapter 2).

The complex situation may, however, be resolved into about a dozen major groups of peoples. By far the largest of these is the *Slav*, comprising over 70 per cent of the total Soviet population. Of the 190 million Slavs, 137 million are Great Russians (almost 53 per cent), closely followed by the Ukrainians (or Little Russians) totalling 42 million. There are also nine million Byelorussians as well as much smaller numbers of Poles, Bulgarians, Czechs and Slovaks. Historically, the Great Russians and the Ukrainians have spread eastwards to form a continuous wedge from European Russia across Siberia, in the process of which they have mixed with and absorbed members of non-Slav peoples, notably Finno-Ugrians in northern European Russia, Tatar peoples in the Volga basin and even Mongols and others in Siberia. In the latter amalgam emerged the *Sibiryak*, proud of the mixed Russian ancestry and well adapted to the harsh environment. The Great Russians have been essentially forest dwellers, whereas the Ukrainians – the 'people of the border' – emerged historically in the steppe confrontation with non-Slav nomads, and among them developed the remarkable socio-military 'fraternity' of the Cossacks, becoming the spearhead of tsarist expansion in the eighteenth and nineteenth centuries. The Byelorussians ('White Russians'), the smallest Eastern Slav group, are also a forest people, cast by history in the role of serfs and peasants variously under Polish, Lithuanian and even Great Russian domination.

The second largest ethnic group comprises over twenty nationalities associated together primarily through language, the Turkic peoples. Over half this group live in Central Asia and the southern Siberian steppe, another quarter live in the Volga basin, while half a million are strewn across the vastness of Siberia, comprising in all about 39 million people, something like 15 per cent of the total Soviet population. In the Volga basin, Turkic groups like the Chuvash and the Tatars have been strongly influenced by Slav settlers and now live scattered among mostly Russian communities. The Turkmen and Uzbeks, largely sedentary oasis cultivators, are much influenced by Persian culture. The 12.5 million Uzbeks are the third most populous people in the USSR. The Azerbaydzhanis in Transcaucasia, numbering over five million, are also a Turkic people extensively Persianised. In contrast, the Kirgiz nomads have Mongol influences. Some of the nomadic groups, notably the Kazakhs, suffered severe losses in early Soviet days through resistance to collectivisation.

The third largest ethnic association yet comprising little more than 2 per cent of the total population is the highly diverse Japhetic population of the Caucasian lands. The major nationality, with a powerfully developed identity, is the Georgian (Gruzinian), numbering well over half the total of this group. The other peoples are small in number, mixed amid peoples from other ethnic associations: for example, in Dagestan there are over thirty different nationalities belonging to several different ethnic associations. With both farming and pastoral traditions, Christian and Moslem culture and other diverse traits, disunity more than any other quality has distinguished the history of these peoples. In the view of Soviet ethnologists, the Armenians (some 4.2 million) are regarded as ethnically distinct. With their own Gregorian Christian faith and a long tradition as merchants, they have played a special role in the historical evolution of both Soviet and other territory.

Just under 2 per cent of Soviet population is formed by the 4 million Finno-Ugrians, diverse and scattered. Historically, these peoples have been pushed ever further northwards into increasingly poorer environments or have been absorbed by the expanding Eastern Slavs. Some of the most important members of this group live in the Volga basin, notably the Mordovs, Udmurts and Mari, but the other major concentration is in the north-west, formed by the Estonians, Karelians and Finns. The most backward of these peoples live in the far north, like the Komi, Komi-Permyak and Saami (Lopari), and least developed are the reindeer-herding Selkups, Nentsy and Khanty-Mansi in northern Siberia. Nearly all have been influenced by contact with Russians. Soviet ethnologists include in the Finno-Ugrian group the 171,000 Magyars living within Soviet Carpathia.

Fifth in order of size are the Baltic peoples, mostly Latvians and Lithuanians, where Polish and German influence has been strong. Their ancient languages have been a vital element in the retention of their national identity, especially as they are among the few peoples of the Soviet Union using the Latin alphabet. Historical influences kept Slav infiltration of their homelands at bay until late in the nineteenth century.

Iranians (just over 1 per cent of the Soviet population) are spread from the Caucasian isthmus to Central Asia, three-quarters belonging to the Tadzhik nationality, a people with relatives in northern Persia. As already noted, the Turkic Azerbaydzhanis are culturally closely associated with the Iranians.

Among other significant ethnic groups are the Moldavians (just over 1 per cent of the total population), closely related to the Rumanians, against whom they have been used unsuccessfully by the Russians as a political lever. Forming under 1 per cent but politically much more important than that suggests are the Jews, whose numbers have fallen steadily. Historically, they come from several different backgrounds, though they have been notably urban dwellers spread right throughout the country, and an attempt to attract them into a separate Jewish national area (located paradoxically in Eastern Siberia) during the 1930s failed. It is perhaps a surprise that the fourteenth largest nationality in the Soviet Union comprises the 1.9 million Germans. Despite wartime 'liquidation' from their settlement areas in European Russia to Siberia (Kulunda Steppe) and Central Asia, they remain remarkably numerous.

The Mongols form less than 0.2 per cent of the total population, and over a quarter belong to the Buddhist Kalmyks of northern Caucasia. The remainder belong to the various Buryat groups of Baykalia. The Tungus-Manchurians (Eveni–Evenki) comprise a mere 0.02 per cent, and the Palaeo-Asiatics (mostly in the far north-east of Siberia), a mere 0.01 per cent. Groups numbering a few thousands are spread across areas greater than Europe (e.g. the 39,000 Eveni–Evenki). Some peoples count their members in hundreds, like the 800 Yukagirs and 500 Aleuts. These are mostly simple hunting, fishing or herding peoples. There are also native to Soviet territory small numbers of Tibeto-Chinese peoples, Koreans, even Greeks and Gypsies.

Of ninety-three nationalities, only twenty-three number more than a million persons each, and of these only seven exceed five million. As already noted, the scene is effectively dominated by the Great Russians (137 million) and the Ukrainians (42 million), whose nearest rivals are 12.5 million Uzbeks and the 9.4 million Byelorussians. With the Slav population comprising almost three-quarters of the total population (the Russians alone 52 per cent), it may truly be said that the Soviet Union is a slavonic wolf masquerading in a sheepskin of multinationalism. Even the next largest group, all the diverse Turkic peoples together, comprise a mere 15 per cent of the total (approximately 10 per cent in the interwar years), but showing little evidence of any sense of union, they constitute an even less effective counterweight to Slav dominance than their numbers suggest. The three Eastern Slav nationalities – the Russians, Ukrainians and Byelorussians – with a little effort can be mutually intelligible and show a much greater degree of cultural similarity and solidarity (the feelings of Pan-Slavism) than found among the other associations, all elements that strengthen their political pre-eminence. Nevertheless, many key peoples among the non-Slavs have shown a much greater rate of natural increase than the Slavs since the Second World War, so that although the latter's position is secure for a long time ahead, there are already signs that a shift in the ethnic balance towards the Turkic people in particular has begun. Perhaps the greatest challenge to Slav predominance could arise from the resurgence of a militant Islam manifested since the late 1970s, a potentially powerful unifying force among many non-Slav nationalities.

Although the Constitution recognises and guarantees the right of nationality, accepting for some groups a distinct territorial-administrative identity (see Chapter 2), the true nature of the 'nationality problem' needs consideration. A traveller in any union republic outside the RSFSR will be struck by the everyday use of the local language including publications and in radio and television and much is also made of local culture. He will, however, also be impressed that notices in public places are in both Russian and the local language and that in bookshops Russian-language works tend to overshadow the local language. He will be in no doubt that the *lingua franca* is Russian and that anybody wanting to make his or her way in Soviet society must have a thoroughly fluent command of it, besides the fact that in all corners of the union republics there are considerable minorities of Russians who appear to hold key positions. Soviet statistics suggest that over three-quarters of the total population speak Russian fluently and hardly no one passes through school without gaining some reasonable knowledge.

With such a wide range of cultural achievements among the nationalities and such a numerical imbalance, how do the less-developed and numerically small peoples survive being swamped by the larger and more advanced groups? The local levels of territorial-administrative national identity (see Chapter 2) were designed for such a situation, where groups could be protected and helped to raise their economic and cultural levels. Nevertheless, all these groups seem to be relentlessly drawn into the Soviet web, and although the authorities have tried to protect some of the lesser-developed groups, they have at the same time tried to spread Soviet ideology among them. It is interesting that the codification of several of their languages in the 1920s was first based on the Latin alphabet, but in Stalin's sovietisation these were changed into a modified Cyrillic alphabet. It is recognised that once members of these peoples go to work in or have close contact with mining and industrial enterprises staffed by Slavs their assimilation is usually quick, while many native women are assimilated through marriage or looser association with Slav workers.

The problem of loss of national identity into the fabric of an essentially Russianised Soviet milieu also faces the more advanced groups with long historical traditions of nationality. Sovietisation has been reluctantly accepted by the Baltic peoples, who for a period between the wars were independent, but there has also been a fierce stance against the regimentation and standardisation of the Soviet milieu by the Ukrainians and the major Caucasian peoples. Concern arises through the continuing immigration of Russians or other outsiders into the various union republics; through the tendency to intermarriage; as well as through the increasing need to use Russian as a *lingua franca* in scientific and administrative affairs. Under such pressure, many people change allegiance to the Russian nationality, officially recorded on the internal passport carried by every Soviet citizen. Despite the reputation of the Slav for successful intermarriage and assimilation of non-Slav peoples, well seen in the Volga lands and in Siberia, there has been resistance to assimilation among the Islamic communities of Central Asia, where fundamental cultural differences between Moslem teachings and Christianity have kept the societies apart, despite the Soviet regime's efficient atheistic policy. The maintenance of any compact national settlement area becomes difficult as increasing mobility of population is demanded by economic development and shifts in the spatial pattern of the economy. The problem was reflected in the 1970 census, when over 21 million Russians and more than 5 million Ukrainians were recorded as living outside their own republics. Moreover, the titular groups in some republics no longer predominated: in Kazakhstan, for example, the titular Kazakhs comprised only 32 per cent of the population, whereas Russians formed almost 43 per cent; in the Kirgiz SSR, the titular group formed less than 44 per cent of the population. In hardly any of the autonomous SSRs do the titular groups predominate.

3.2 **Geographical distribution of population** (see Figure 3.3)

One of the greatest problems in the Soviet Union is its distribution of population in relation to the spatial pattern of economic activity. Figure 3.3 outlines the main features of population distribution, dominated by the striking triangular-shaped belt of territory with over ten persons per km^2 whose base lies between the Baltic and the Black Sea and the apex in Western Siberia. The main part of this

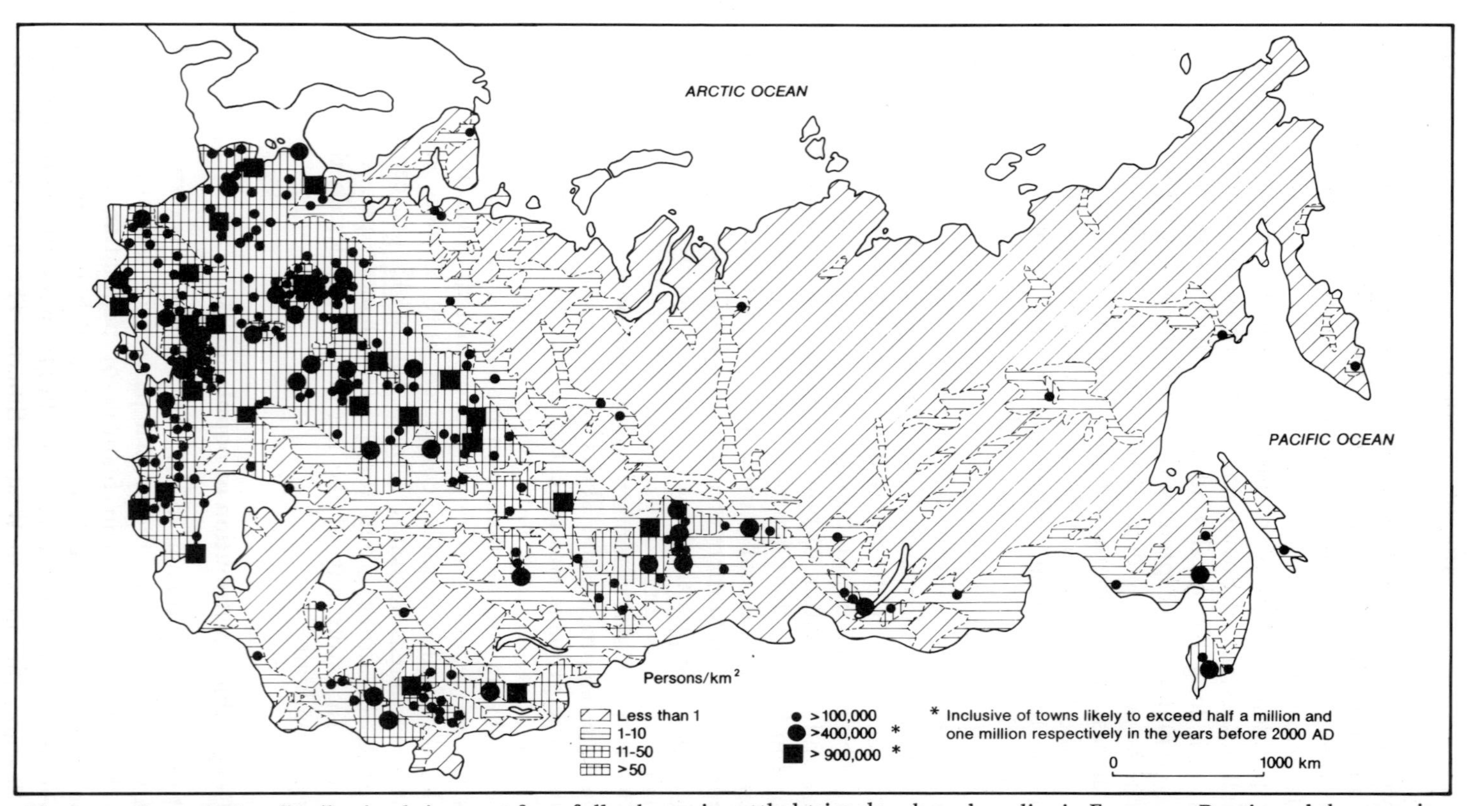

The map of population distribution brings out forcefully the main settled triangle whose base lies in European Russia and the apex in Western Siberia.

Source: updated from *Atlas SSSR* (Moscow, 1969).

Figure 3.3 *Population distribution in the USSR*

triangle, the one-sixth of the Soviet Union that lies west of the Volga and south of Leningrad, contains two-thirds of the total population. Outside this triangle are only two other notable concentrations of people – in the fertile oasis lands of Central Asia and in the Transcaucasian republics. Less than a quarter of the national area has more than ten persons per km^2, whereas almost 70 per cent has less than five persons per km^2.

Seven out of ten Soviet citizens consequently live in European Russia, notably south of Leningrad and west of the Volga. North of 60°N, though more thickly settled than corresponding parts of Siberia, densities fall sharply, with large patches of the extreme north with less than one person per km^2, but, as common through the Soviet Union, rivers and railways form ribbons of settlement. People in the better country are mainly farmers or forestry workers, while the native peoples in the poorer country are reindeer herders. The population total is boosted by the scattered mining towns of the Kola Peninsula and northern Karelia, the towns of the Ukhta oilfield and the Pechora coalfield (Vorkuta, 101,000) as well as Arkhangelsk (387,000) with Kholmogory and coastal ports like ice-free Murmansk (388,000).

The north-west of European Russia is dominated by Leningrad (4.6 million) and its nearby towns. This is better settled farming country (mostly with well over ten persons per km^2) but also has large tracts of forest. Also well settled is the glaciated country of the Baltic coast, though some sparse tracts occur in north-west Estonia, and many principal settlements are seaports like Tallin (436,000) and Riga (843,000), with its many nearby pleasant resorts. Inland, extending into the areas of Russian and Byelorussian settlement, the landscape of villages amid fields and forest is typical, but settlements tend to cluster on drier ridges, particularly evident in the position of towns such as Vilnius (492,000), Minsk (1.3 million) and Smolensk (305,000). Along the Byelorussian–Ukrainian border in the exceedingly wet Polesye and Pripyat marshes densities drop below ten persons per km^2 and occasional small hamlets lie on the drier strips of sand.

The central part of the European plain is dominated by Moscow (8 million), the greatest of all Soviet towns, and its associated industrial communities that form a striking cluster. Locally, with so much urbanisation, densities may reach over 100 persons per km^2, though, to the south and south-west, wet country, like the Meshchera, is more thinly peopled. North-east and east of Moscow the countryside is dotted by many quite large textile towns, attracted by the humidity, such as Shchelkovo (100,000), Noginsk (120,000) and Pavlovskiy Posad (69,000). Growth throughout this part of Russia has been chiefly in the towns, with only most modest rural increases at best and even declining patches in some poorer parts like the remoter Valday Hills, with powerful migration from the countryside since 1945.

The drier open country south of Moscow has densities of around fifty persons per km^2, exceeded in some more favoured parts. This agriculturally better country with good soils witnessed considerable population increase in the last century resulting eventually in acute rural overpopulation. Between the wars it became a countryside from which migration was particularly strong, but industrial towns like Voronezh (796,000), Lipetsk (405,000), Tula (518,000) and Kaluga (270,000), let alone many smaller ones, have become industrial growth points. With development of the large iron-ore resources around Kursk (383,000), this countryside is likely to see the foundation of further new mining and industrial settlements.

Population density of the Ukraine averages over eighty persons per km^2, but despite the densely settled rural population in many districts the proportion of urban population is well below the average for the Soviet Union. The well-settled nature of the Ukraine is tellingly reflected by the fact that it contains almost 20 per cent of Soviet population and yet only 2.7 per cent of the total territory. The high rural densities, often well over 100 persons per km^2, are found especially in a broad belt of country extending from the northern foot of the Carpathians eastwards towards Kiev, with good soils and comparatively adequate and reliable rainfall. This belt also projects southwestwards into Moldavia. The high densities in the extreme west of the Ukraine and in western Moldavia are partially a legacy of incorporation of rurally overpopulated countryside since the Second World War from which there has been a vigorous migration. Elsewhere, to the south and east of the Dnepr, extensive tracts of reasonable farming country with more than fifty persons per km^2 occur, where rolling fields stretch away to the horizon with settlements lying sheltered in straggling lines along gullies and valleys where there is water. The moisture problem becomes acute towards the Black Sea coast, so that densities fall appreciably as farming becomes extensive in form, dipping well below ten persons per km^2 in parts like the dry steppe that projects into the northern Crimea. The Crimean mountains are sparsely peopled by pastoralists, but sheltered garden-like valleys on both the northern and southern slopes are closely settled. The southern slope along the Black Sea is dotted by some of the most famous Soviet holiday resorts, such as Yalta (81,000) and Massandra (pop. unavailable), and there are also good harbours (e.g. Sevastopol (308,000) and Feodosiya (78,000)).

The Ukraine is nevertheless a part of the USSR with some highly industrialised districts, of which most striking are the towns of the western Donbass coalfield that include Donetsk (1,032,000) and nearby Makeyevka (439,000), Voroshilovgrad (469,000) and Gorlovka (337,000). Another cluster lies in the great bend of the Dnepr, including Dnepropetrovsk (1,083,000), Dneprodzherzhinsk (253,000), Zaporozhye (799,000) and the mining town of Krivoy Rog (657,000) with its satellite communities. Further up the Dnepr lies Kiev (2.19 million), the Ukrainian capital, and to the east of the river, Kharkov (1.46 million), once the capital, both large engineering towns. Several important harbour towns lie along the dry Black Sea coast, notably Odessa (1.057 million), Nikolayev (449,000) and Kherson (324,000). On the Azov coast, Zhdanov (507,000) is a major iron and steel town.

In the RSFSR the eastern Donbass and the lower Don form another cluster of important industrial towns, including the port of Rostov (946,000). The western part of the Northern Caucasian region is formed by the well-settled farming country of the Kuban, with mostly more than twenty-five persons per km^2, though the marshy lagoons on the coast have a sparse population. Well-settled country lies along the northern footslope of the Great Caucasus, where there is not only good farming country but also settlements associated with the oil industry. Towards the east of the isthmus, however, particularly in the dry steppe of the Kuma basin, farming gives way to pastoralism, with densities in places well under five persons per km^2.

The Volga lands, like much of the south of European Russia, were a pioneer frontier in the eighteenth century and filled rapidly with settlers last century, a settlement that still continues, notably with the stimulus of oilfield development and associated industries. The river is lined by important towns like Gorkiy (1.36 million), Kazan (1,002,000), Ulyanovsk (473,000), Kuybyshev (1.2 million), Saratov (864,000) and Volgograd (939,000), with Astrakhan (465,000) the port at the delta. Densities are generally higher along the west bank, because the low eastern bank is liable to extensive floods in spring. Some of the greatest densities occur on the outlier of better soils south of Kazan in the homelands of the Chuvash and Mordovs. On the lower Kama a well-developed agriculture and new industries have brought densities of over fifty persons per km^2. Below Saratov densities on the east bank drop away quickly towards the dry south-east, but on the lower reaches the availability of water and the many fertile islands in the braided lower course raise densities to over twenty-five persons per km^2 in the most favoured spots. It has been, however, the middle reaches around Kuybyshev where the most striking increase in population has taken place since the Second World War as a result of massive industrial development.

The Ural region is larger in area than England and Wales together, containing about 17 million people, some 6 per cent of Soviet population. Peppered with manufacturing and mining towns, it is regarded essentially as an industrial region and was a major growth area between the wars that underwent further considerable increase after 1945, characterised by a strong inward migration. Growth has been concentrated in the urban industrial communities, which show a marked distribution along the mineral deposits of the eastern slope, but there has also been a considerable growth on the west in the Belaya and Kama basins. The central section of the mountains, crossed by numerous easy routes, shows the greatest settlement densities, with many small industrial towns, but the main route focus is Sverdlovsk, now a city of over a million people, and other clusters of towns lie around Chelyabinsk (1.04 million) on the east and Perm (1,008,000) on the west. Most settlements line the valley floors along railways and roads, with almost empty forest or modestly settled farmland between. Densities generally are around ten persons per km^2 but rise in the more favoured country to well over twenty-five persons per km^2. Oil wealth and good farming country make the western footslope appear more densely peopled than the gentle eastern dip into the limitless West Siberian lowland. The Ural ranges north of 60°N are extremely thinly peopled, like the drier southern part, though in the south the population is boosted by industrial towns like Novo-Troitsk (96,000), Orsk (252,000) and Orenburg (471,000) and the better-settled country along the Ural valley.

Siberia comprises almost 57 per cent of the area of the Soviet Union and yet, despite vigorous encouragement to settlers since last century, contains just under 10 per cent of the total population. Distribution shows a clear contrast between the near-empty northlands and the straggling line of settlement along the Trans-Siberian railway, which carried the main inflow of settlers, both farmers and workers for mines and industries. After the Revolution, when the Soviet Union was essentially a closed economic system, development of Siberia was ruthlessly pursued, bringing a strong immigration (including large numbers of 'involuntary' settlers), but the rapid growth has not been maintained, so that the share of total population peaked in the 1950s and has fallen back slightly in the 1970s. The gross underpopulation of Siberia remains one of the most intractable of all Soviet problems. Despite all kinds of incentives to migrants, the harsh environment constantly proves unattractive and there is a high turnover of labour as people drift to more salubrious parts of the country. As living standards rise, the difficulty becomes greater to draw people to these remoter areas, and on present performance it looks as though the official projection of 60 million people in Siberia by the year 2000 will fall short by up to 40 per cent.

The most striking feature of Siberia is the long straggling lines of settlement along the widely meshed communications system, whether roads, railways or (particularly) rivers. The belt along the Trans-Siberian railway is notably prominent, with densities mostly around ten persons per km^2 and patches rising to around twenty-five persons per km^2. On the west the belt is broad, the area of original sedentary settlement late in the nineteenth century extending south into the newer farming lands of northern Kazakhstan. The main towns are all railway centres, with Omsk (1.02 million), the major junction, as the largest, but one of the most thickly peopled areas lies around the million-person city of Novosibirsk, one of the principal Soviet scientific and cultural centres, effectively the Siberian 'capital', and south along the Ob valley into the Kuznets Basin. The Kuzbass coalfield is a cluster of big industrial towns like Kemerovo (478,000) and Novokuznetsk (545,000); in all, some eight towns have each more than 50,000 people. The belt of settlement along the railway continues eastwards towards Lake Baykal through Krasnoyarsk (807,000), Cheremkhovo (75,000) and Irkutsk (561,000).

East of Lake Baykal, principal towns are Ulan Ude (305,000), in the Selenga basin, and Chita (308,000) in the open country of the Shilka valley, around which the main population clusters occur. Transbaykalia has well-peopled valleys between forested uplands, where relatively few people live, and densities are generally less than in Western Siberia. The grassland basins support considerable indigenous population, notably the Mongol Buryats. Apart from mining and railway settlements, population density in the Amur valley falls from densities of around ten persons per km^2 on the west to well under ten persons per km^2 towards Khabarovsk. Here quite a few settlements exist to supply railway labour, while settlement has been encouraged for strategic reasons along the Manchurian frontier. Once Khabarovsk (538,000) is reached there is better-settled country, where densities rise to over twenty-five persons per km^2 in some places along the Ussuri valley and even above this in the fertile lands around Lake Khanka and Vladivostok (558,000), chief port of the Far East. A narrow band of settlement extends north from Khabarovsk along the Amur valley, with the main settlement around the steel town of Komsomolsk (269,000). Patches of settlement also occur along the Sikhote Alin coast, where several ports have been developed in recent times (e.g. Soviet Harbour and Nakhodka). The rough mountain terrain along the southern Siberian frontier in the Altay and Sayan ranges is virtually uninhabited except for the valleys and basins, where mining communities have been founded in Soviet times.

The great northlands are marked by their emptiness, with population density everywhere under one person per km^2, apart from small clusters in favoured spots or along the river valleys. Most of these settlements are concerned with mining, transport or the infrastructure for the simple economies of the sparse and largely nomadic native peoples. The effect of the great rivers carrying vast quantities of warm water north to the Arctic gives an important local moderating influence on climate along them that attracts settlement. The Siberian northlands have, however, some of the highest percentages of urban population in the whole USSR, since in Magadan *oblast*, for example, 376,000 of the 478,000 people (79 per cent) are urban. There are also some remarkably large urban settlements, like the mining town of Norilsk with 182,000 inhabitants, the port of Magadan (124,000) or the Yakut capital, Yakutsk (155,000), itself located in a more thickly settled patch of better country in the middle Lena valley. The building of the great Baykal–Amur trunk railway (to be completed by 1983) will attract renewed settlement in the potentially rich Aldan and Vitim plateaus, having perhaps the same effect as the building of the hydro-electric barrage at Bratsk (219,000) in the 1950s.

Soviet Central Asia shows great contrasts between the crowded fertile oases and the empty inhospitable desert or modestly settled poor steppe country, while there is also lonely high mountain country broken by well-peopled valleys and basins. The key to settlement distribution is water – long-established indigenous communities exist where it is naturally available, whereas there are newly settled lands wherever it has been provided in recent time by artificial irrigation. The steppe lands of northern Kazakhstan, the scene of considerable colonisation in Soviet times from the rural countryside of European Russia, carry densities of up to ten persons per km^2. Other significant clusters of settlement occur wherever there has been mining or other economic development, most notably around the coalmining town of Karaganda (577,000), while the nearby steel town of Temir-Tau has risen from 5,000 in 1939 to 215,000 in 1980. An unusual settlement is the 'cosmodrome' at Baykonur, scene of Soviet space research, while there has been rapid growth in the 1970s of oil-working communities in the arid Mangyshlak Peninsula. Elsewhere in the steppe and semi-desert, there is a thin scatter of indigenous shepherds, with the most sparsely settled areas in the Ust-Urt plateau and the true desert of the Kara-

Kum and Kyzyl-Kum as well as the infamous Bet-Pak-Dala or Hunger Steppe.

The dry lands are broken, however, by long riverine oases like the Amu and Syr Darya valleys. Here densities of fifty to seventy-five persons per km^2 are quite common, while in places densities of over 100 persons per km^2 are found, as in the rich farming area of the ancient delta of Khorezm near Khiva and Tashauz. Along the mountain foot lies a broad belt of loessic materials and silt whose soils are extremely rich when watered and within the mountains are similar fertile basins. Densities from thirty to well over 100 persons per km^2 typify the irrigated lands, and even on the dry farming areas there may be ten to twenty persons per km^2. Around the edge of the Fergana Basin on delta fans of mountain streams, rich farming supports up to 200–300 persons per km^2, reaching over 400 persons per km^2 near Andizhan, whereas the centre of the basin in contrast is a sandy wilderness. There are many significant towns in and around the main irrigated lands, like Tashkent (1.82 million), Alma Ata (928,000), Andizhan (233,000) or Bukhara (188,000).

The mountainous border of Soviet Central Asia is, apart from its moister basins and valleys, sparsely settled, mostly by indigenous pastoralists gathered into various forms of collective organisation. Density falls rapidly with altitude, though some well-settled basins occur at considerable elevations, while there is still transhumance up and down the slopes with the seasons. As elsewhere, mining and industrial communities have been developed since the Revolution, peopled mostly by immigrant Slavs.

This characteristic pattern of mountain settlement also occurs in the Transcaucasian mountains, where, expectedly, altitude and aspect play an important role in the settlement pattern and density. There is also much movement up and down the slopes with the seasons, but the majority of people live in some form of closely nucleated permanent settlement for most of the year. South of the Great Caucasus as well as in the mountains live distinctive peoples with their own long traditions in settlement forms. In the warm humid lowlands and valleys, notably in western Georgia, densities of forty to fifty persons per km^2 are typical, though the most favourable parts have densities well in excess of 100 persons per km^2 gathered into nucleated villages and towns. The Caucasian Black Sea littoral is also well populated, for its warm sunny climate gives it an attraction for both intensive fruit cultivation and for recreation in seaside resorts. Eastern Caucasia, in both the mountains and the lowlands, is much drier and consequently less thickly settled, unless there are special economic attractions like the oilfields, notably those of the Apsheron Peninsula and now the offshore workings in the

Caspian Sea. In many parts of the eastern lowlands densities fall to below twenty-five persons per km^2 and some patches of the Kura lowlands are even more thinly settled. Baku has grown to a million-person city on the strength of its petroleum wealth, but it is closely rivalled by the attractive Georgian capital, Tbilisi (1.08 million). The lower parts of the Armenian plateau support twenty-five to fifty persons per km^2, but the highest and bleakest areas carry less than ten persons per km^2, with scattered indigenous shepherds and farmers. The luxuriant and sheltered valleys below the plateau are thickly peopled, particularly where there is irrigation and near the principal towns of Yerevan (1,036,000) and Leninakan (210,000). Although it claims only about a twentieth of Soviet population, Transcaucasia has been one of the fastest-growing parts of the Soviet Union in recent times.

3.3 The role of migration in the population pattern

The contemporary pattern of population distribution is the product of a long historical process of migration, which in some ways may be seen principally as a domino effect of the eastwards and southwards movement of the Eastern Slavs from their area of primary dispersal in the plains of European Russia. The spread of the Eastern Slavs into their vast *Lebensraum* has been one of the least recognised colonisations by Europeans, but it accounts substantially for the present ethnic distribution pattern of the Soviet Union.

The early diaspora of the Slavs from their cradle somewhere between the Carpathians and the Pripyat marshes carried them into the eastern forests, the wooded steppe and the fringe of the true steppe. Fierce nomad incursions into the steppe from within Asia forced their retreat on many occasions deep into the forests, as (for example) between the tenth and thirteenth centuries, but their northward and eastward expansion into the forests brought them into contact with other peoples. In the more easily defensible terrain of the Baltic littoral moraines, earlier peoples held their own against the insurgent Slavs, but the Finno-Ugrian tribes of the forests, less well organised and resolute, were either pushed into more inhospitable areas by the advancing front of Slav penetration or remained in patches of poor terrain behind it, like the groups still found in the Valday Hills.

From the eleventh century onwards the Slavs began their penetration across the Volga and the Ural, even though the southern gateways to Siberia were closed by hostile nomads, but progress along the great rivers and across the swamps of Western Siberia was relatively slow and open to only the hardiest. Defeat of the nomads holding the easier southern entrances in the mid-sixteenth century loosed a rapid spread, mostly of trappers and adventurers, so that within a century a Slav presence on the Pacific coast had been established, with numbers increasingly re-enforced by those banished from the tsar's European domains.

By the sixteenth century declining nomad power in the middle and lower Volga had also encouraged a spread of Slavs southwards, but it was effectively not until the major conquests of the eighteenth century that the so-called 'New Russia' began to fill with settlers in a now unstemmable tide towards the Black Sea and eventually into northern Caucasia. Although through the nineteenth century tsarist power spread in Central Asia, Slav colonisation was numerically modest, limited to a few special areas, and most immigrants were military and notably Cossacks and administrative personnel.

The emancipation of the serfs in 1861 released a new thrust for living space by migration from overpopulated farm lands of European Russia, so that by the end of the century the new railway into Siberia was carrying a tide of settlement eastwards. Over five million people, overwhelmingly Slavs, moved into Siberia in the last century and its population more than doubled by 1914, when annual migration was running at over 700,000, though not everybody could settle permanently in its harshness and many returned, but nevertheless between 1903 and 1913 the net gain was reputedly 4 million.

The Soviet period has generated a massive pattern of movement, and the frontier of settlement has continued to expand outwards through a shift in emphasis from sedentary farming to the development of mining and industry, so consequently everywhere there has been a powerful drift from the countryside into the mushrooming towns. At the same time, these movements have generated a considerable interregional migration. Between the wars economic development of the eastern regions attracted considerable migration into the Ural, Siberia and the Far East, but compared with the pre-1914 period this was markedly to work in new mining and industrial communities rather than as farm settlers. Although these regions have continued to attract immigrants since the Second World War, there has been a change to a vigorous settlement campaign in northern Kazakhstan for both mining and farming development and in central Kazakhstan for primarily mining and industrial projects, notably around the Karaganda coalfield. Since the mid-1960s there is evidence of efforts to attract people to new industries in the long-neglected westernmost parts of European Russia, especially in Byelorussia and the western Ukraine.

The Soviet period has marked a ruthless attempt to people the harsher lands of the interior, notably the immense northlands of

Siberia. Under Stalin, when the prison population was claimed to be around ten million, much labour for development projects in these inhospitable regions came through various measures of duress, and large numbers of prisoners (not all political detainees) were shipped to camps deep in Siberia, but even free citizens were commonly put in a position where they could not refuse a job in these distant corners. Many of these workers, having worked out their contract, or prisoners having served their term, found it impossible to return to their home districts, either because no transport was available or simply because a residence permit was refused. Although the vast prison population of Stalin's day appears to have been greatly reduced, coercion to settle in the least attractive places still seems to be in practice, possibly because in campaigns for free settlement the numbers of volunteers never seem to have matched requirements. Indeed, a dilemma of Soviet inner colonisation remains the problem of the large turnover of migrants, for even in good years almost as many people leave Siberia, for example, as enter. Settlers commonly quickly drift back to the more salubrious parts of the Soviet Union, and even within Siberia there is a marked drift of people from the harsher north to the more pleasant conditions of towns in the railway zone. The bait for the potential settler in the harsher parts is nowadays the incentive of better pay and other material concessions. However, labour in the northern settlements increasingly comprises shift workers, flown in from their homes in towns along the Trans-Siberian railway for two or three weeks at a time.

Although we may regard migration in both tsarist and Soviet times as dominantly a Slav movement (after all, Slavs form more than three-quarters of the total population), it has also played a vital historical role among non-Slav peoples in creating the ethnic quilt of today. For example, the Finno-Ugric peoples appear historically to have migrated from somewhere in the Altay region, while many of the non-Slav peoples of the Volga basin are comparatively late incomers, and the southern steppes have attracted a long procession of immigrant peoples of many different ethnic affiliations from within Asia. As late as the eighteenth century the Buddhist Kalmyks, a Mongol people, entered the steppes of northern Caucasia from deep inside Asia. Some groups seem to have migrated into relatively unattractive areas under pressure from more virile and dynamic competitors: by their culture and language, the Yakuts, for example, appear to have been pushed into the middle Lena basin from more southerly grassland homelands.

The tsars used migration as a political weapon and settled Cossacks and other reliable elements in an imbricate pattern amid unreliable conquered peoples, while they also laid down rigorous conditions for

Jewish settlement, containing it in a *Pale* in western Russia. The almost unlimited power of great landlords to move their serfs about as they saw fit also contributed to mixing the ethnic pattern, while the tsars' readiness to use foreigners introduced a cosmopolitan element to many of the larger towns. On other occasions they attempted to dragoon whole Central Asian peoples into military service, breaking them up into small units stationed far from each other and their homeland. Early this century Chinese settlers around Lake Khanka and in the Ussuri valley were ruthlessly driven out into Chinese Manchuria.

Resettlement in small groups was commonly meted out in early Soviet times to break up troublesome peoples or communities. Such a fate befell, for example, the Kazakhs when they resisted collectivisation in the late 1920s and between the wars their numbers fell by over a quarter. Large numbers of non-Slav peoples were drawn into Stalin's forced labour battalions. During the Second World War some nationalities who were accused, rightly or wrongly, of siding with the invading Germans were 'liquidated', notably the Volga Germans, the Kalmyks, the Crimean Tatars and some small Caucasian mountain peoples. The Germans were transported into Siberia and Central Asia, but the Kalmyks simply officially ceased to exist, even though remaining in their own settlement areas. In the late 1950s these peoples, except the Crimean Tatars, were in various measure 'reinstated'. Potentially unreliable Chinese and Koreans in the Far East were forcibly resettled around the Aral Sea in 1940. Considerable numbers of people regarded as politically unreliable in the annexed western territories and the Baltic republics were moved into the interior by various forms of duress after 1945. In nearly all non-Slav areas in Siberia and in Central Asia, industrial and mining development has been possible only by using skilled labour, itself recruitable principally among the Slav peoples of European Russia. Consequently, the Slav presence among the other peoples has been strengthened as considerable Russian and Ukrainian minorities have multiplied in the economic growth points. Although this process is continuing, peoples like the Uzbeks and Kazakhs in Central Asia and the Georgians, Azerbaydzhanis and Armenians in Transcaucasia, and even the Buryats in Baykalia, have been developing cadres of skilled personnel increasingly able to provide some of the high-quality labour needed in their own territories.

The population dilemma remains a major constraint on the achievement of the economic and political objectives of Soviet policy. It is really a problem of having too few people for planned developments, people that are all too often at the wrong place at the wrong time. The difficulties of the spatial distribution of population resources are made worse by not having mastered the demographic problem of how to regulate the age–sex structure of that population to fit the nation's infrastructural requirements, notably the rising shortage of labour and a growing proportion of old people. These difficulties, bad enough in themselves, are exacerbated by the fact that 'Soviet people' comprise an extremely mixed ethnic collection, each 'nationality' with its own cultural and demographic traits and its own latent aspirations not always conforming to the Politburo's objectives. All these compound to create potentially serious internal political-geographical undertones to population management and inner colonisation. We may look in horror at both tsarist and Soviet authorities for their apparent ruthlessness towards people as well as at the limits to freedom of movement imposed by internal passports. Yet there seems an inevitability to such a pattern of population management in face of the challenges posed in the environmental milieu of that one-sixth of the earth's surface.

3.4 Where to follow up this chapter

The classical historical study of Soviet population remains Lorimer, F., *The Population of the Soviet Union: History and Prospects* (League of Nations, Geneva, 1946). Also useful in Dewdney, J. C., *Population Geography* in the 'Soviet Union' series (Hicks Smith, Wellington, New Zealand, 1969). There are also two chapters by Lydolph, P. E., on Soviet population geography in Trewartha, G. T. (ed.), *The More Developed Realm – A Geography of its Population* (Pergamon, Oxford, 1978), and a special issue of *GeoJournal* (1980 Supplementary Issue 1) on the 1979 Soviet census.

Also significant are Kovalev, S. A. and Kovalskaya, N. Ya., *Geografiya Naseleniya SSSR* (Moscow, 1980), Pokshishevskiy, V. V., *Geografiya Naseleniya SSSR* (Moscow, 1971), and Rukavishnikov, V. O., *Naseleniye Goroda* (Moscow, 1980).

Demographic aspects are covered in Valentey, D. I. *et al.*, *Demograficheskaya Situatsiya v SSSR* (Moscow, 1976).

On ethnic problems, useful are Allworth, E. (ed.), *Soviet Nationality Problems* (Columbia University Press, New York, 1977), Goldhaven, E. (ed.), *Ethnic Minorities in the Soviet Union* (Praeger, New York, 1968), Kozlov, V. I., *Nationalnosti SSSR* (Moscow, 1975). There are also Bromley, Y. V., *Sovremenniye etnicheskiye Protesy v SSSR* (Moscow, 1977), and Katz, Z. (ed.), *Handbook of Major Soviet Nationalities* (Free Press, New York, 1975).

4

Settlements – Where Soviet People Live

In the days of the tsars the Russian empire was a land of villages and hamlets, while some non-Slav people moved seasonally with the needs of hunting or herding. The Soviet authorities, for both political and economic motives, set in motion a vast programme of urbanisation, so that between the First World War and the early 1960s the proportion of town dwellers rose from under a fifth to just about half. By 1980 just over 63 per cent of the population was living in urban settlements. The formidable growth meant an increase in town dwellers from 28.5 to 169.3 million. As a consequence of the powerful migration from country to town to achieve this change, many country districts have seen their population numbers stagnate or even decline. By the year 2000 it is expected that the proportion of town dwellers will possibly exceed 75 per cent. In some remoter areas where there has been extensive development of mining and industry without corresponding agricultural settlement, the proportion of urban population exceeds 70 per cent.

4.1 The village

For many of the peoples of the Soviet Union the village or hamlet has been the traditional element of settlement. Where the older villages remain little changed, they show considerable regional diversity. Wherever Slav settlers went they took with them their village form and house type, especially in Siberia, though there were many variations in the size of the village and in the detail of the houses. In tsarist times a village was distinguished according to whether it had a church (*selo*) or was without (*derevnya*), but in Soviet times the differentiation is between the villages that act as seats of collective or state farms and those that do not.

In northern European Russia villages are small, widely scattered on drier sandy mounds or river terraces, and seldom comprise more than a dozen houses. In central European Russia, the north-west and Byelorussia villages of several dozen houses are common, but house types vary from one district to another in size and richness of appearance. Especially poor and meagre are the villages of the wet lands such as the Polesye. The most common layout everywhere is the straggling street village, a simple line of houses on both sides of a broad, unpaved track. In the south the steppe villages seek shelter from bitter winter winds and commonly lie along valleys and erosion gulleys in long lines, where one village merges into the next, each distinguished by its own pond. The newer-settled areas of the Ukrainian and Caucasian steppe are marked by large chessboard villages, best developed in the Cossack communities of the Kuban. Villages in south-western Siberia are also large and prosperous-looking, with generous use of space.

Villages in the Baltic republics tend to be small, comprising about twenty houses or so. Land reforms before the Soviet takeover tended to scatter farmsteads, but nucleation has been encouraged again by collectivisation. Karelia is a land of small villages and hamlets, with large houses and barns. Irregularly shaped villages near springs or streams mark Tatar settlement in the Volga lands. In the Caucasian mountains tightly nucleated villages of irregular plan, comprising usually flat-roofed stone houses, stand on easily defended slopes, in some districts marked by strange guard towers. In the foothills and lowlands villages are large, comprising distinctive houses with verandahs, while some Caucasian houses show Turkish influence in stone basements with overhanging upper floors in wood. Central Asian villages are large and closely nucleated, surrounded traditionally by a low wall. Their flat-roofed courtyard mud houses with no outside windows crowd along narrow alleys.

Although in Soviet times effort has been made to ease the life of the poorer nomad and hunting peoples, particularly by encouraging them to become sedentary, many still follow their traditional types of settlement. In Central Asia and in Siberia nomads live for at least part of the year in traditional tents like the *yurt*, even though they are supplied with necessities from collective- or state-farm centres. In arctic and subarctic Siberia winter houses are still built partly below ground from wood, stone or even sods, and are deserted in summer for tents or huts raised on low stilts.

The effort to modernise agriculture has included considerable plans to restructure rural settlement. Marxist-Leninist ideology has laid particular stress on equality of material and social conditions between the peasants in the countryside and the industrial workers in the towns. Nevertheless, living conditions in most villages have lagged well behind the towns, notably in the facilities available. Much effort has been devoted to improving housing, and the standard factory-made house based on the style of the Russian *izba* has been replacing the various traditional regional types. Other important improvements needed have been in rural water supply and in roads, though provision of electricity has been greatly expanded. At the same time, there has been a need for adequate schools, shops and recreational opportunities.

There have been long-standing Soviet proposals to replace individual farmsteads (as in the Baltic republics) and the small villages and hamlets sited away from roads or with other locational disadvantages. As early as the 1930s, in the northern Caucasian grainlands, attempts to develop town-like farm settlements had been made. Khrushchev pushed the development of the agricultural town, the *agrogorod*, in the new wheatlands of Kazakhstan. Nevertheless, some urban elements, like three- or four-storey blocks of flats, have begun to appear in the larger villages, where effort has been devoted to separating housing from the other farm buildings. In the early 1960s plans were formulated to develop a little over a sixth of all villages as key settlement points in the countryside; the remainder, it appears, would be allowed to stagnate or might even be removed. In Byelorussia, for example, about 27,000 villages, mostly with about ten or so families, were to be reduced to 5,000 major rural settlements. Implementation of the plans, widely unpopular among the rural population for both political and social reasons, have generated much controversy and there has been disagreement on how fast and how far they should be pursued. Even so, considerable reduction in the number of rural settlements through reorganisation has been marked in Eastern Siberia and the Far East, while change has been least in the most thickly settled parts of European Russia, as in the Baltic republics and in Transcaucasia.

4.2 The town

The Soviet authorities, for both planning and ideological reasons, regard the town as the most vital element in settlement. As the 'proletarian base', it plays an important role as the favoured milieu for the evolution of a communist society in the internal political geography of the Soviet Union. According to Marxist-Leninist theorists, it is the focal point from which to disseminate ideology to a not uncommonly reluctant countryside and must therefore reflect the achievements of the regime by being 'national in character, socialist in content'.

Table 3.1 above (p. 47) reflects the growth in the numbers of urban dwellers since the Revolution. Soviet sources suggest that since the Second World War about 60 per cent of this growth has been from migration from the countryside, about 20 per cent has come from natural increase in the urban population and probably about 17 per cent from the reclassification of rural settlements as 'settlements of town type'. Migration has usually been from the villages into comparatively nearby towns, particularly in European Russia. Elsewhere, in the eastern regions for example, a considerable proportion of the migrants come from far away. Siberian towns have gained most settlers from towns in European Russia. People move to towns for betterment in employment or education, and the higher living standards of town life compared with the countryside have been a powerful attraction. A considerable proportion of migrants into towns are single people or young married couples, though in some remote and inhospitable places there has been difficulty in attracting sufficient women to build a balanced community. Many arctic towns and workers' settlements have originally been peopled by migrants under duress, principally politically unreliable elements, an unfortunate tradition dating back to the eighteenth century when some embryonic industrial settlements in the Ural arose from the enforced migration of serfs, while quite a few Siberian towns in tsarist times had begun as prison settlements.

Although the birth and death rates in towns have been lower than in the countryside, the difference between these rates has been less than in the countryside, so that consequently the rate of natural increase in towns has been lower than that in rural areas. Families in the countryside have been generally larger than in towns, despite there being a higher proportion of women in the child-bearing age groups in the latter. Although the rate of natural increase has been high in the rural areas, the drift to the towns has drawn away so many young people that many rural areas have consequently shown substantial decline in their population, cumulatively marking an overall decline in rural population at the national scale. Increase in town population has been not just in existing communities but also in new towns or in places elevated to town status as soon as they reached the necessary threshold for such designation, while the boundaries of many towns have been substantially extended to include adjacent settlements.

Figure 4.1 reflects the regional variation in the proportion of urban population in the territorial-administrative districts. The

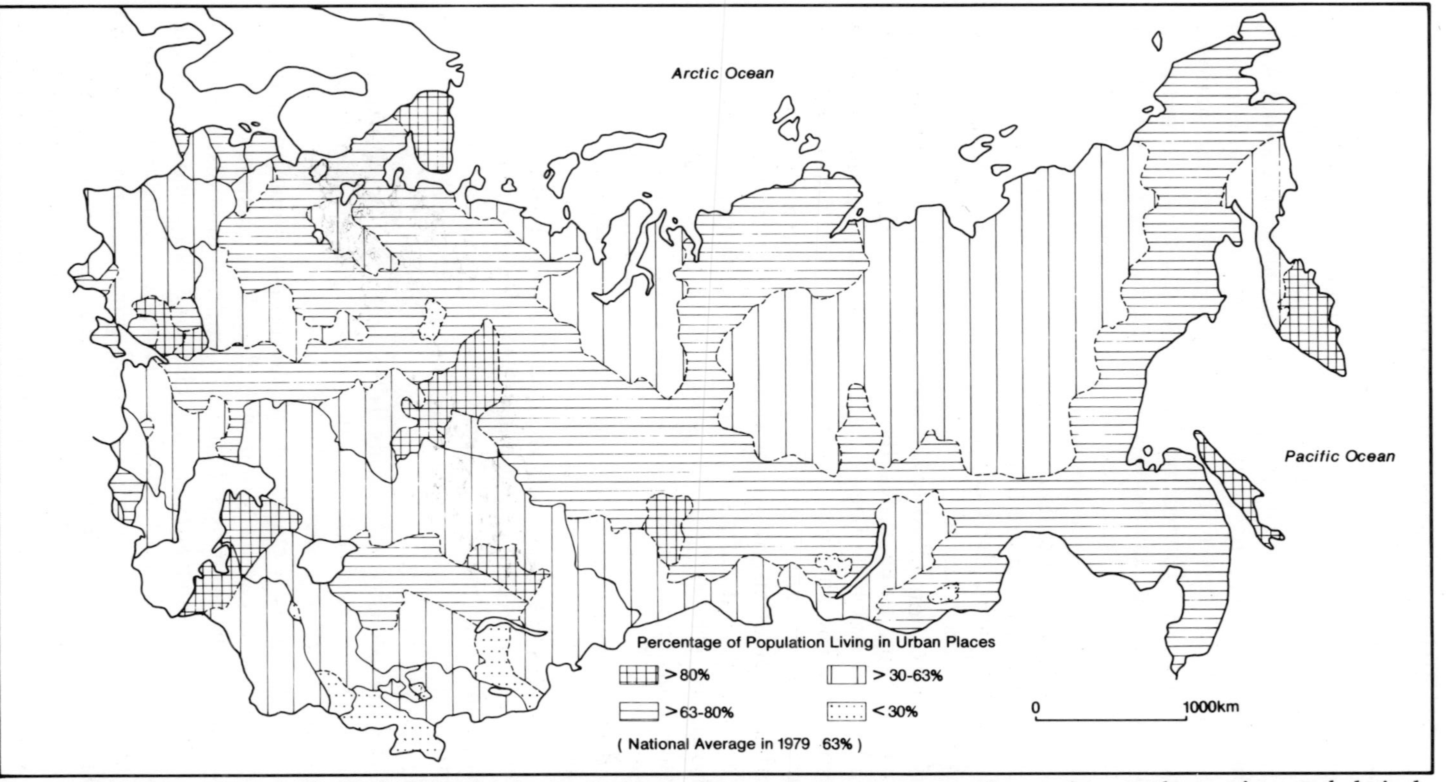

The map illustrates the remarkably high levels of urbanisation in the more remote eastern regions, where settlement is overwhelmingly in a few scattered mining or industrial communities.

Source: *Narodnoye Khozyaystvo SSSR* (Moscow, 1979).

Figure 4.1 *The distribution of Soviet urban population*

proportion of urban dwellers is generally quite a bit below the national average (63 per cent) in the older-established farming districts, as, for example, in Western and Central European Russia, northern Caucasia and the Ukraine, or the oasis areas of Central Asia, all comparatively densely settled lands by Soviet standards. In contrast, urban dwellers form proportions well above the national average in the lands of newer settlements where emphasis has been on mining and industry, characteristic of much of Siberia and the Far East, the European North, the Ural region and some parts of Kazakhstan. In such regions town population is often largely Slav, and rural population is non-Slav. Urban population is also above the national average in the more industrialised parts of European Russia, as in the Volga basin, the Central Industrial Region and the industrialised parts of the Ukraine.

There are now over 5,900 towns and settlements of town type in the Soviet Union, three times the number in 1926. Well over a thousand new towns have been created in one way or another since the Revolution. The absolute growth of urban population continues at a rapid pace, though its annual percentage increase slows as the total urban population grows. In the post-war years the towns of between 100,000 and 250,000 persons, most favoured by planners, have shown the greatest growth; growth rates for the larger classes progressively decrease, as they do for the smallest towns. Nevertheless, the number of towns with more than a million people has increased from two in 1926 to three by 1939 and to twenty in 1980, while several more are currently near the critical threshold. All the signs suggest that Soviet aims to promote the growth of the medium-sized towns at the expense of the very largest have not been notably effective and the 'million cities' seem to possess a peculiar momentum of their own, despite a rigorous official control of the granting of residence permits.

The regional density of towns is expectedly uneven, bearing, however, a general relationship to the over-all pattern of population distribution, and though north of 60°N towns are few and scattered, they nevertheless play a major role in the total regional population. In European Russia can be found several striking clusters of towns. The most northerly comprises Leningrad and its immediate satellite towns. There are some fifty towns clustered immediately around Moscow itself, as well as a marked concentration of industrial towns to the east, extending towards the small group around Gorkiy on the Volga, besides some clusters to the south towards the Donbass, which is itself one of the largest urban constellations in the whole USSR. Elsewhere in European Russia there is otherwise a remarkably evenly spaced scatter, densest in the Ukraine and quite impressive in the Baltic republics; nevertheless, a noticeable tendency is seen for towns to occur in strings along the main lines of movement, once again reflecting the peculiarly dominant role of transport. This comes out strongly in the Volga valley and also on the major routeway along the northern footslope of the Caucasus, repeated also along the east–west 'trough' between the Great and Little Caucasus. A further remarkable line of towns lies roughly along the 60° E meridian arising from the rapidly urbanised mining and heavy-industry settlements in the Ural between the wars. Again, however, from a detailed map of town distribution, we can also pick up the east–west routes across the Ural by the pattern of towns, and the linear pattern of towns is perhaps nowhere more forcefully developed than along the Trans-Siberian railway. In Central Asia the oases along the mountain foot or along the nilotic rivers, with their many small native towns and town-like settlements, tend again to emphasise the linear nature of distribution. Some early Soviet work on urban geography investigated specifically the patterns of towns in urban clusters and we may consider whether the *pattern* of spatial distribution of towns becomes significantly more important in such a vast and empty territory like the Soviet Union compared with, say, the thickly settled countries of Western Europe, though in south Germany it was certainly used by Christaller in his theoretical work on the urban hierarchy.

The definition and subsequent status of urban settlements depends on essentially economic criteria (Table 4.1). Urban status is in each instance related to a minimum adult population, a minimum proportion of whom must be engaged in industry or related occupations, though requirements vary somewhat between the republics (see Table 4.1), being usually least exacting in the less-developed republics. Once a settlement has reached the status of a town, its further progress up the urban hierarchy is related to the level of local government body to which it is subordinated, whether to the *rayon, oblast, kray* or to the republic (usually only for the republic capital). Moscow has a special status of its own. If fortune is adverse, however, a settlement may be downgraded, like several small market towns in western territories annexed in 1945 unable to develop a truly industrial structure and consequently reduced to village status. Any 'workers' settlement', the settlement of town type and first step on the urban hierarchy, may be downgraded or may even be dissolved and disappear should the original reason for its establishment, like a mine or major construction project, be abandoned. Rather special in this category are the health-resort settlements, found throughout the country, though areas like Caucasia and the Crimea show strong concentrations. Near the largest towns are also

Table 4.1 *Settlements of town type – criteria for definition in selected republics*

Republic	Employment structure (% workers in industrial sector and member of their families)			Minimum total population		
	Workers' settlements	Settlements of town type	Towns	Workers' settlements	Settlements of town type	Towns
RSFSR	Not less than 85%	Non-existent	Not less than 85%	3,000	Non-existent	12,000
Ukraine	Preferably majority	Over 60% (50% for rayon centre)	Preferably majority	500	2,000	10,000
Georgia	Non-existent	Not less than 75%	Not less than 75%	Non-existent	2,000	5,000
Moldavia	Preferably majority	Not less than 70% (not less than 60% for rayon centres)	Preferably majority	500	2,000	10,000

Source: Khorev, B., *Gorodskiye Poseleniya SSSR* (Moscow, 1968).

'dormitory' *dacha* settlements, regarded as a form of urban development. The advantages of town life from the ideological viewpoint encouraged attempts to try to bring it to the agricultural population, with early efforts to establish agricultural 'towns' in northern Caucasia. The true *agrogorod*, as already noted, came with the Virgin Lands scheme of the early 1950s, where farm workers live in a town-like community, commuting to their work in the fields each day.

Soviet towns fall into two basic categories: those existing at the time of the Revolution, and those subsequently founded on green-field sites. In the former group, alongside the elements arising from the implementation of Marxist-Leninist ideas of the town, there are also strong historical influences, not always of Russian origin. For example, many towns of the Baltic republics show in their morphology and their buildings powerful German influence, while Lvov still bears a strong visual impression of its growth under the Habsburg empire, just as Uzhgorod, Mukachevo and Chernovtsy also show the impress of their former allegiances. Some of these older towns incorporated after 1945 that were extensively damaged, like Kaliningrad (former German Königsberg), have, however, been rebuilt in a characteristic Soviet style. In Transcaucasia and in Central Asia towns of extremely ancient origin with powerful Persian or Turkish influences, incorporated into nineteenth-century Russia, have still retained their form, so that the 'Russian town' remains distinct.

The traditional Russian town lies in a dry and defendable position, typically on the high right bank or terrace above a river, usually at a route intersection or the end of a portage. The original nucleus is commonly a walled town (*gorod*), containing a fort (*kreml*) or even a monastery or cathedral as well, reflecting the ties between church and state in Old Russia, seen in Moscow, Kiev, Smolensk, Chernigov and Pereyaslavl. Around this enclosed town it was common for artisans and others to settle, ultimately themselves being enclosed within a wall, as illustrated by Moscow. At Novgorod the enclosure with the administrative and church functions lay on a steep bluff on the western bank of the Volkhov, while across the river was an enclosure containing the merchants' quarters. Some towns became the venue of large seasonal fairs held outside their walls. The most important was at Nizhniy Novgorod (*now* Gorkiy), where ground for the fairs lay across the Oka opposite the town, itself on a low terrace below where its *kreml* stood on a commanding bluff.

The new towns established as Russian power spread southwards and eastwards began life mostly as fortified posts. The typical Siberian towns became palisaded, wooden trading posts set commonly on a bluff by a river where there was sufficient good ground for cultivation to feed their inhabitants, and where routes intersected at portages, fords or ferries. Even today many Siberian towns remain essentially built of wood – wooden houses, wooden sidewalks and once brightly painted but now dilapidated wooden churches, while the administrative buildings are colour-washed stucco on a wooden frame. Because so many intellectuals were exiled to them in the nineteenth century, Siberian towns became known for their excellent museums, theatres and libraries.

The intensified economic development of Soviet times has led to the founding of over 3,000 new settlements, all on greenfield sites, scattered across the country, with many in the remoter parts of Siberia or the poorer steppe and semi-desert of Central Asia, notably Kazakhstan. Quite a number of new towns have been founded in European Russia, particularly in the Ukraine and even not far outside Moscow, besides the eastern slope of the Ural and in the Volga basin. Many of these new towns have grown to considerable size, like Magnitogorsk in the Ural founded in 1929, now with 410,000 people, Karaganda, founded in 1928 as the centre of the coalfield in Kazakhstan, with 577,000, and nearby the steel town of Temir Tau, now a town of 215,000 but with only 5,000 people in 1939. In Siberia several towns existing before the Revolution have shown boom characteristics – Novosibirsk, a mere village when the railway reached it in the mid-1890s, is now the Siberian 'capital' with 1.3 million people; Norilsk, a shanty town when founded in 1935, reached 14,000 people in 1939 and 182,000 in 1979; Angarsk founded only in 1948 and yet by 1979 a town of 241,000 inhabitants; whereas the iron and steel town of Komsomolsk founded in 1928 on the banks of the Amur had risen to 269,000 people in 1979.

Some of the new towns founded even as late as the 1960s have been built principally in wood in the old tradition, as (for example) at Divnogorsk, the construction centre for the large Yenisey hydroelectric barrage near Krasnoyarsk. In the far north Aykhal and Deputatskiy have been planned to be under great plastic roofs, with buildings hanging from the structure rather than with conventional foundations. This avoids the difficulty of heaving in the permafrost and also allows an artificially more moderate environment to be developed. Nevertheless, problems of water supply and sewage disposal are acute wherever there is extensive permafrost. In contrast, in the arid lands of Central Asia the major problem of settlement is water supply. Artesian sources are most suitable, but in some instances canals have been built from nearby rivers, despite excessive loss from evaporation. The new oil town of Shevchenko on the arid Mangyshlak Peninsula uses atomic power to desalinate local supplies, whereas Krasnovodsk further south receives water by tanker from across the Caspian Sea. In such arid environments artificially provided greenery (minimum 25m^2 per inhabitant) gives shade.

The ideal of Soviet planners is the 'town of socialist realism', where the concept of 'national in form, socialist in content' is adequately expressed. Commonly associated with a major industrial development, the focus of the town is its main industrial plant, though there is usually a belt of greenery or a reservoir for industrial water between the town and the works. The town is designed and laid out in neighbourhood units, whose size depends to some measure on the ultimate planned population of the town itself. Each unit is complete with its own supporting infrastructure, the aim being to reduce 'unproductive' urban transport to a minimum by keeping people as far as possible in their own neighbourhood. As in most Soviet towns, a striking feature for the Western visitor is the fewness of the shops and other services, provided not on a competitive basis but on a strict ratio to the number of inhabitants. The town core contains most administrative functions, specialised shops and central services, but a vigorous effort is made to avoid the 'dead heart' (so much criticised in Western towns) by providing residential accommodation right into the centre, which comprises imposing public buildings such as theatres, cinemas and higher-education institutions. The main railway station and a large square with broad approach streets suitable for mass public demonstration form the usual focal points. Planners in Eastern Europe have criticised Soviet towns for being conceived with too great a compactness, supposedly intended to give the so-called 'big-city effect', resulting in monotony and overcrowding of buildings that are too high and line overwide streets, while it is claimed the layouts are commonly too geometric and focus too forcefully on the main industrial plant.

In the older towns the same elements are also found, though woven into and sometimes overshadowed by the pre-Soviet framework. Buildings range from traditional styles in old churches, monasteries or the *kreml* and town walls to various modern styles. Wooden buildings remain common, even in the great cities, but the mass of buildings are in undistinguished brick or neglected stucco. Public buildings and blocks of flats range from the flamboyant, even bizarre, Stalinist style to more formal and austere functional designs, either dating from the late 1920s or from a recent return in the 1960s. Away from the imposing centre, building standards drop quickly, with much shoddy, neglected property along poorly paved streets, and the fewness of shops is offset by small kiosks and mobile stalls, while small factories and workshops are scattered amid a mixture of tenements. Trams still form a significant element in town transport along the main thoroughfares, though there are also buses and many taxis, with a dusty square away from the centre used as a station for country buses. Outside the great cities, in the smaller town, street traffic is generally light. A busy place in all towns is the collective farmers' 'free market', where not only food but also other goods are traded, while allotment gardens are a recurrently common feature in every town. In all Soviet towns one impression in common is how much greenery they possess, either from tree-lined streets and gardens or in parks, for each has its own large Park of Rest and Culture. The outer suburbs are usually a mixture of small wooden houses in large fenced gardens along unpaved streets or new housing developments in large blocks of flats, again with the familiar vegetable gardens. Although space is used generously, the transition from town to country is usually sharp. The low priority received by housing for too long left a serious backlog, made worse by the continued vigorous growth in the number of town dwellers and the pressure to improve living standards. Towns with the worst housing deficits imposed rigorous restrictions on residence permits, but this was not completely effective. It was not until the early 1960s that vast building programmes for rehousing and additional accommodation were instituted. The extensive allotment gardens reflect the wish of recent migrants from rural areas to grow their own vegetables and keep a few rabbits as well as the need of established townsfolk to help their own food supply, a valuable contribution to the feeding of the growing urban population in an economy that seems incapable of providing an adequate distributional system for most consumer goods.

There has been a long discussion in the Soviet Union on the optimum size of the town so that the best amenities can be provided in relation to a satisfactory return on investment and to keeping running costs to the minimum. In the early 1930s, in the prevailing mood of gigantomania, proposals were mostly for multi-million population towns, though nowadays it is usually accepted that the optimum population is between 250,000 and 500,000 people, varying according to local circumstance. A view of the mid-1950s that no town should exceed 400,000 people seems to have been dropped in the face of the inability to control the growth of the largest towns as the number of 'million cities' has risen from three in 1959 to twenty in 1980. In both Minsk and Kiev, for example, populations exceeded a million inhabitants well in advance of the planners' forecasts, while estimates of ultimate populations in new towns have invariably proved too low, so that the chemicals town of Volzhskiy on the Volga, originally planned for 50,000, has risen to over 200,000 and the ultimate size is now projected as at least 300,000.

Trends in settlement in Soviet territory exhibit little difference from those elsewhere – the ongoing growth in urban population and the change in long-established patterns of rural settlement. Perhaps the most important element in this changing pattern is the favour shown, notably on ideological grounds, towards towns. The world-wide drift from country to town is an equally marked feature throughout the Soviet Union, though it is perhaps worth noting that the proportion of rural dwellers (37 per cent) remains conspicuously higher than in the major industrial nations of the capitalist world.

4.3 Where to follow up this chapter

Soviet urban problems are examined at length in Bater, J. H., *The Soviet City: Ideal and Reality* (Arnold, London, 1980), and by French, R. E. and Hamilton, F. E. I. (eds), *The Socialist City: Special Structure and Urban Policy* (Wiley, New York, 1979), as well as Harris, C. D., *Cities of the Soviet Union* (Chicago University Press, 1970). The Soviet view is found in Khorev, B. S., *Problemy Gorodov* (Moscow, 1971), and *Gorodskiye Poseleniya SSSR* (Moscow, 1968). Moscow is covered by Saushkin, Yu. G., *Moskva – geograficheskaya Kharakteristika* (Moscow, 1964), and by Hamilton, F. E. I., *The Moscow City Region*, in 'Problem Regions of Europe' series (Oxford University Press, 1976).

On the village, a useful source is Kovalev, S. A. and Kovalskaya, N. Ya., *Geografiya Naseleniya SSSR* (Moscow, 1980). See also Maynard, J., *The Russian Peasant and other Studies* (Gollancz, London, 1942), and Robinson, G. T., *Rural Russia under the Old Regime* (Macmillan, New York, 1932).

5

Soviet Agriculture

As late as 1959 nearly one-third of the employed persons in the Soviet Union were engaged in some sort of farming; even twenty years later, over a fifth of the labour force still remained in this sector. Agriculture has been the cinderella of the Soviet economy, and the inability to solve in acceptable ideological terms some of its most crucial problems remains a serious constraint on development in other sectors. Despite its importance to an economy that has sought for so long to be as self-contained as possible, agriculture has been for many Soviet people a symbol of the past, with the position of farming and farmers adversely influenced by social attitudes moulded by Marxist-Leninist dogmata based essentially on an urban-industrial philosophy. Not only was the countryside starved of necessary investment, but it was expected to feed the whole population and at the same time provide labour for the new industries and inhabitants for the new towns. In the prevailing atmosphere the most able and progressive country people sought their fortunes in the new urban-industrial milieu. As the urban population grew and demanded better living standards, the task of the farmers became more difficult, with agricultural produce sorely needed at home frequently sold abroad to finance purchases not of farm machinery but of specialised industrial equipment. There has even been a planning dilemma between planting food crops or crops with industrial application.

As economic development has been pushed into remoter and less hospitable regions, a growing disparity between location of food production and population has arisen. Yet with heavy burdens on the transport system, attempts have been made to develop food production in areas with markedly sub-optimal conditions at consequently high cost, but this may be seen as part of the dilemma of distribution.

Two ways of increasing food production may be taken: first, yield may be raised per unit of existing agricultural area; or second, the agricultural area may be extended without a necessary raising of unit yield. The first is usually most readily achieved on the best agricultural land, while the second commonly offers most possibility in areas where only limited farming activity has so far developed. These problems are also influenced by the over-all pattern of physical conditions under which the Soviet farmer has to work that are over the greater part of Soviet territory relatively hostile compared with the natural environment for farming in Western Europe or much of North America.

A little over a fifth of the land area of the Soviet Union is used agriculturally compared with over a half in the USA. Because of the more northerly latitude of the Soviet Union and its position extending deep into the continental interior of Eurasia, vast areas are quite unsuited to arable farming and are even marginal for pastoralism. Through most of Siberia, forestry or herding, with small patches of arable or meadow, are typical, while over the poor steppe or the mountainous parts of Central Asia and Transcaucasia, only low-density pastoralism is possible. Despite the extensive plains, winter cold or summer drought make successful crop-raising hazardous over the greater part of the arable belt, while large spreads of soils of mediocre fertility give extremely modest yield without considerable use of artificial fertilisers. Nevertheless, many large areas could be made more productive if adequate drainage or irrigation were provided and techniques improved.

In a country composed essentially of lowlands, climate and soil, rather than slope, aspect or altitude, are the principal factors affecting both current farming and the extension of the farmed area. The great distance from the sea of much of the country means a generally low precipitation and an important relationship between it and evaporation. Over the north of the country, low temperatures, moderate precipitation and little evaporation on the gentle plains make water-logging common and good drainage essential for successful farming. In contrast, in the south, hot summers and a rather low precipitation (occurring mostly in thundery spring and summer showers) mean evaporation greatly exceeds precipitation and there is much loss of rainfall by run-off on parched ground. Consequently, the successful farmer must carefully husband his moisture, a task exacerbated by the general unreliability of the rainfall (Figure 5.1). Eastwards, increasing continentality reduces the growing season and the frost-free period, with rapid changes between winter and summer that shorten the time for spring sowing and the autumn harvest. Nevertheless, in the high latitudes the long days of summer tend to

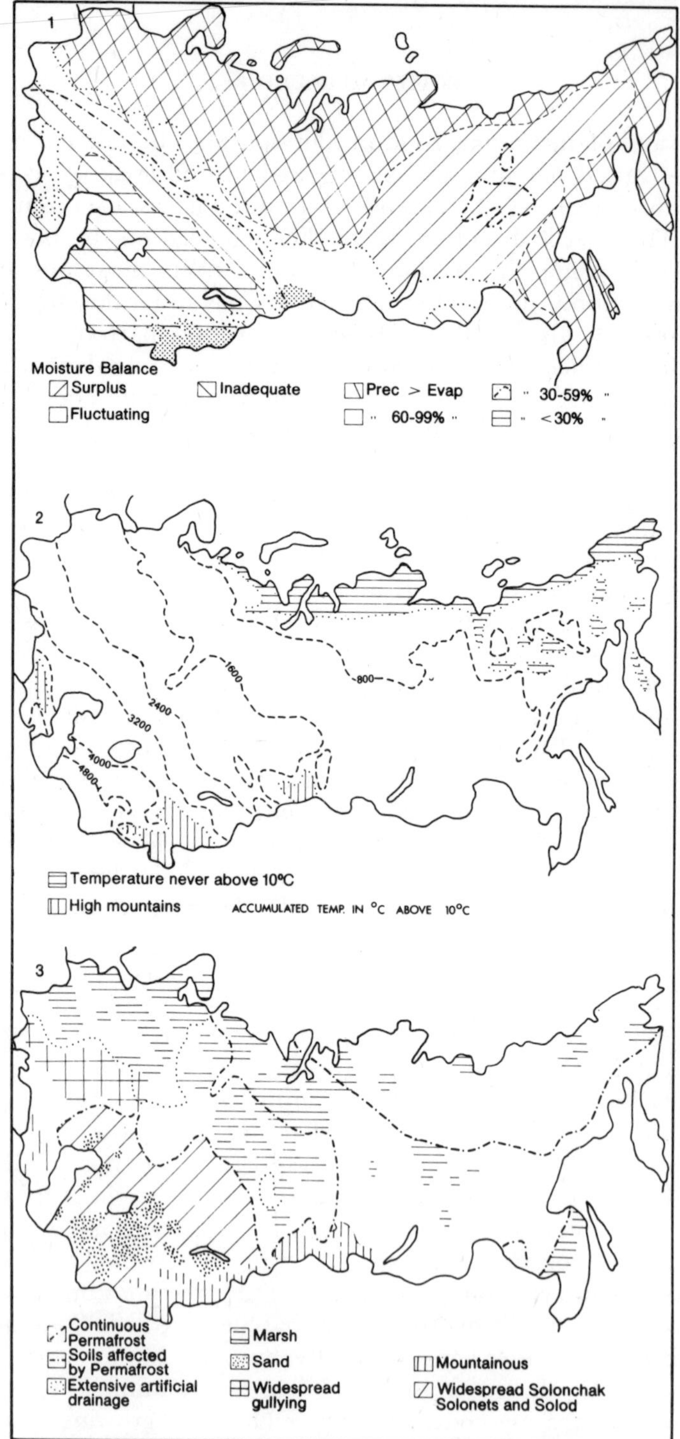

1. Moisture 2. Temperature 3. Terrain

The maps suggest the difficulties that face the Soviet farmer. Once again the most suitable conditions occur in the triangular area whose base lies in European Russia and the apex in Western Siberia.

Sources: various Soviet atlases.

Figure 5.1 *The physical conditions of farming*

offset some of these effects, but even so more than half the country has a growing season of less than a hundred days. In the great winter cold, particularly in the eastern regions, much of the country has only a thin snow cover, easily blown away by the typical high winds, while late spring or early autumn frosts make farming operations hazardous. There is also the additional constraint of permafrost in otherwise promising parts of Siberia. Although the steppelands are ideal grain country, cultivation is problematical because of aridity and unreliable precipitation made worse by scorching summer winds, so that soil erosion and the threat of dust-bowl formation are an ever-present menace, conditions which tend to reduce grain yield but intensify labour requirements by the need to dry farm. Irrigation, an attractive solution to the moisture problem, must be planned and operated with great caution if devastating salt-pan formation is to be avoided.

Indeed, climatic conditions alone rule out agriculture over about a third of the Soviet Union – about 15 per cent accounted for by the tundra, 10 per cent by the desert and the balance coming from mountainous districts. Almost a half of the country's area is covered by boreal coniferous forest, much underlain by permafrost, where agriculture is worth while in only a few favoured patches. Under a tenth of the forest belt is reputedly used for arable or meadow.

Even where conditions are otherwise suitable, poor and mediocre soils hamper farming. The best soil is by far the black earth (*chernozem*), but even including much degraded types it covers no more than 12 per cent of Soviet territory, two-thirds located in southern European Russia. Siberian black earth is poorer in quality, with about a quarter of its extent not suited to farming because of salt-pan formation. Nevertheless, unused reserves of *chernozem* lie in Siberia and along the Kazakh border. The black earth retains its remarkable fertility despite heavy cropping and limited application of fertiliser, but is highly friable and careless husbandry has already lost much good land to sheet erosion and gullying. This soil stores winter moisture for a rapid and luxuriant growth in spring, but crop yield is influenced by availability of summer moisture when the dry *sukhovey* blows.

Over half the area of the Soviet Union has soils of the forest podzol type, cold, acid and poor in humus, but with suitable fertiliser application and also proper drainage they can be made agriculturally useful. The most cultivable of these soils occur in the southern forest belt and overlap into the wooded steppe, while in Eastern Siberia there are stretches of low-acid podzol. The extensiveness of this belt suggests that the greatest opportunities for extending the cultivated area lie within it.

The chestnut, grey and brown soils of the drier steppe and semi-desert extend widely across the south of the Soviet Union. They need great care in cultivation, for large tracts have undesirable salts in the upper layers that make watering or irrigation difficult. If too damp, however, they become sticky and unworkable, whereas dry they can be iron hard. The areas affected by salt (*solonchak, solonets* or *solod*, depending on the nature of the accumulation) are generally considered not worth reclamation. Along the southern edge of the desert belt, rivers have brought down rich silts and muds, while deep loessic deposits have built up around the mountain foot; both can be extremely fertile when properly watered and form the extremely productive oasis soils of Central Asia.

Soviet farming inherited a grim legacy from the past. Serfdom had spread and been strengthened during the seventeenth and eighteenth centuries, but unlike medieval Europe it was used not to produce just for its own stomach but for the market. At the time of the emancipation in 1861, well over half the agricultural labour force was comprised of serfs, even after the early nineteenth-century freeing of peasants in the Baltic provinces, a partial reform in Russian Poland and many free peasants in Moldavia. In the poorer agricultural provinces of the north, where tied labour produced relatively little, most landlords accepted a tax payment (*obrok*) from their serfs, who were otherwise free to earn what they could by whatever means they chose. In the richer black-earth lands the landlords preferred serfs to give labour (*barshchina*), fixed at three days a week but commonly exceeded, since handsome profits could be made in grain and other surpluses shipped to the north where there was usually a food deficit. In the north and centre much land was also owned by the church and crown, both happy to receive tax payments, like the many absentee landlords. Nevertheless, the trend everywhere was generally for money payments to replace labour service. Peasant society was dominated by the village commune, the *mir*, that fixed the redistribution of strips of land and the rotation of crops. It was encouraged by the state and the landlords since it relieved them of much tiresome administration without giving the serfs any independence. However, with the growth of the commercial market for farm produce in the nineteenth century, the more far-sighted landlords began to see that serf labour could not provide the efficiency agriculture increasingly needed, whereas industrialists wanted to end it in order to swell the labour available for their new factories.

Emancipation in 1861 did not end the land problem, for peasants were left without capital and guidance to improve their techniques and output. Holdings acquired at emancipation varied considerably in size between districts, but were generally too small to give real economic security and independence. Many took 'pauper holdings' (one-sixth of the normal area) to avoid paying redemption duties (anyway too high for most peasants to pay), but even so most quickly fell into debt, while in the end the land titles remained vested in the unpopular *mir*. Indebted peasants sold out to more successful and prosperous ones or to the landlords, notably in the more fertile south, where the holdings were in any case generally smaller than before emancipation, because the landlords had become interested in their full share of the valuable land.

Emancipated peasants began to leave the overcrowded districts, particularly in the northern black-earth lands, to move into towns or into Siberia. The estates were usually without capital to replace the former cheap serf labour with machinery, particularly in the north, but generally the landlords had managed to keep the better ground that could have made balanced peasant holdings. Even so, some landlords unable to face the new situation sold out and retired permanently to the towns. Although numbers leaving the countryside swelled, the land hunger remained, because the rural population continued to increase faster than allotments of land could be made to it, while output remained low, since the *mir* still made the partition of the old three-field rotations, and the peasants, too poor to buy fertiliser, had to maintain a larger area than necessary of fallow each year.

The Stolypin reforms of 1906 generated by growing unrest were meant to end land hunger by ending the authority of the *mir*, by consolidating holdings, and by encouragement to people to leave the countryside to make more plots available. By 1916, however, only a fifth of peasant families had managed to consolidate the whole or part of their holdings and fewer still had managed to leave the village to live among their fields. Some 65 per cent of all peasants were poor and dependent on secondary occupations and a mere 15 per cent were truly prosperous farmers. Most peasants had about the same amount of land as their counterparts in France or Germany, but because of inefficient methods and yield output was less than a third. Had the social and economic strains of the First World War not come, however, Stolypin's reforms, given time, would probably have been sufficient to create a reasonably satisfactory pattern of farming at that date.

It looked briefly as though the Bolsheviks might tolerate small independent peasant farmers, but they quickly favoured a return to communes and a collective type of farming. With the *mir* still active and cultivation of strip fields common over large areas, such a move was not difficult. Peasants seized estates but seldom had the experience to run them, while many estate factors and more able peasants who could have helped had been killed or had fled. Looting and war had destroyed much livestock and equipment, and fields went out of cultivation because peasants held more land than they could manage. Disruption of marketing, confiscations and a lack of manufactured goods to buy discouraged peasants from selling to the market, and with hiring labour forbidden subsistence farming predominated. By 1918 the towns were starving and a hot, dry summer in 1921, with a disastrous crop failure in the Ukraine, Ural and Volga, resulted in a famine in which reputedly 10 million people died.

The New Economic Policy of 1922 sought to avoid disaster by giving peasants an incentive to produce and market through a Rural Code that recognised their ownership of stock, equipment and produce. A dramatic improvement took place until another poor harvest in 1927 was used as an excuse to introduce collectivism again. The result was a fall in livestock numbers, because peasants slaughtered their animals believing the state would provide better ones, and crop yield declined as there was inadequate animal manure and artificial fertiliser supply did not expand.

Collectivisation was an ideological choice and had to be forced through against stubborn resistance from peasants. Many farm managers and committee members were thus chosen with political reliability counting for more than a knowledge of the land and how to farm. It is not surprising that collectives were mostly dismal failures. Farm collectivisation did, however, provide labour for the factories as disgruntled peasants left the countryside, while it also swelled the army of forced labourers as opponents to it were taken into custody. Despite the opposition and the poor performance, by 1940 well over 90 per cent of all peasants were collectivised.

With all priority given in the immediate post-war years to industrial rehabilitation, it was well into the 1950s before the loss of animals and machinery could be made good and the capacity of agricultural engineering and fertiliser production brought up to the requisite levels. Matters were not helped in Stalin's day by acceptance of some strange scientific ideas about farming and grotesquely grandiose plans to change nature itself. Under Khrushchev it was seen that improved living standards and greater incentives for industrial workers could only be achieved through better food supplies resulting from an improved performance from agriculture. Effort was thus made to improve rural living conditions to hold the most able people on the land, to raise yield on existing cultivated ground and to bring new land into use. A key was to be more artificial fertilisers, but this required expansion and modernisation of the chemicals industry. Investment was made in drainage of wet lands in the north and irrigation and watering of dry lands in the south. Although the peculiar genetic ideas of men like Lysenko were quietly dropped, effort was made to develop new plant strains to give better yield and to be cultivable over a wider range of conditions. Particular attention was given to improving the quality of livestock, though this could only be achieved by controlling peasant animals that roamed freely among the farms' own stock.

The most grandiose project was the great ploughing campaign of the 'Virgin Lands', a broad strip of country extending from northern Caucasia across southern Siberia and northern Kazakhstan to the foot of the Altay. This vast scheme, initiated originally under Stalin's rule, illustrates several of the environmental problems faced when trying to expand the cultivated area. The attempt was to establish grainlands on virgin and long fallow soil, most black earth and related forms, in the fringe of the wooded steppe and true steppe, where almost 41 million ha were ploughed between 1954 and 1960. This campaign followed earlier ones in 1928–32 and 1940–4, both of which had helped extend the cultivated area in the Siberian–Kazakh border country. The main constraint on successful cultivation is the low and unreliable precipitation, varying from 300–380mm in the north to a critical 250mm in the south, and in at least one year in five a drought may ruin the crop. Wind is also a serious problem across the open steppe, blowing away the light winter snow cover and the friable topsoil, while thundery showers on the parched ground in summer wash away much good soil. In summer evaporation is intense and scorching dry winds can ruin the crop at harvest time. Among other problems was the need to create an adequate infrastructure of settlement and transport, while difficulty was encountered in attracting settlers to pioneer colonisation. The initial years were successful, with excellent harvests that encouraged the ploughing of more ground, but the 1960s saw both bad years and complete failures, so that in the south considerable areas were ultimately allowed to revert to grassland.

The disappointing results of the Virgin Lands scheme turned interest towards improving land in the European forest belt through the more and better use of fertiliser, better field drainage (over great areas even simple drainage is absent), or just by clearing stones, useful in themselves for filling underground drains. In the Polesye, around Lake Ilmen and in the Oka basin, drainage has been used to establish meadowland rather than arable. It is now appreciated that investment in land amelioration and improvement in the forest zone, where yield is more stable, may be more rewarding than in the more uncertain conditions of the still uncultivated steppe. In the steppe, however, where conditions of natural grassland are the true home of the grains, the main effort has been to assure a regular supply of moisture, with several large hydro-electric barrages on the Dnepr, Volga and other rivers designed specifically to provide water for irrigation. A most ambitious project of this type is the North Crimean Canal, leading from the Kakhovka Barrage on the Dnepr to the dry steppe of the peninsula. There are similar projects in Northern Caucasia on the Kuban, Terek and Kuma rivers, but the Manych system begun in the 1930s still remains incomplete.

With the possibility of growing cotton and crops not otherwise cultivable in the Soviet Union, effort has been concentrated in Central Asia on extending the farmed area by better irrigation. Of the simple systems that existed before the Revolution, most have been radically reconstructed and large new schemes are usually linked to hydro-electricity generation, like the Fergana valley. The Turkmen republic has been supplied with water drawn from the Amu Darya along the Karakum Canal, begun in the late 1950s and still being extended westwards. Well over 10 million ha are now irrigated throughout the Soviet Union, but the potential for which water could be available is estimated at about 30 million ha. Grandiose schemes have been proposed to divert water from northern rivers southwards, including a colossal project to top up the falling levels of the Aral and Caspian Seas by using long-deserted natural drainage channels. By this means it is claimed some 80 million ha could be irrigated. Even in the semi-desert and poor steppe artesian water supplies are being tapped for small areas of cultivation or for live-stock and in the Kzyl Kum desert some thirty-three 'artificial' oases are to be developed by this means to give over 10,000 ha of irrigated land. Irrigated land in Central Asia accounts, however, for only 5 per cent of the total area, though it carries over 60 per cent of the sown area.

The clear desire of farm organisation is to bring the primarily urban-industrial philosophy of Marxism-Leninism to the countryside, despite its inappropriateness to the vagaries of the natural world. On the collective farm (*kolkhoz*) ownership of land, equipment and stock is vested in the farm, and although an individual is free to leave, he may not retain any land, even if he previously owned it. The produce is sold through state marketing agencies at agreed prices and the income divided among members (in terms of 'labour days', agreed norms of daily work) after paying obligations. Collective farmers receive some payment in money and some in kind. Originally machinery was hired from Machine and Tractor Stations, but these were abolished in 1958 and turned into repair stations, and collectives now hold their own machinery. The most important settlements in the countryside are in fact the central villages of the collectives, which have schools, shops, workshops, veterinary units and other services. The main inadequacies of life on most collective farms have been the unpaved roads and the lack of piped water and electricity, but much effort was made in the 1960s to remedy this, and by 1970 virtually every collective had at least an electricity supply, while substantial improvements in housing were being achieved.

The trend has been to amalgamate collectives into larger units and at the same time encourage specialisation, commonly followed by conversion into a state farm (*sovkhoz*) which has a more factory-like organisation and usually more machinery and fertilisers at its disposal. There is a manager and all employees are paid like factory workers. Originally the state farm was set up in areas of agricultural colonisation and development where local settlement did not lend itself to collectivisation, but it was also used to set the pace of farming for local collectives where their performance was poor. The state farms have also been experimental centres, used to introduce new techniques or crops and most are anyway fairly highly specialised. In this way during the 1930s several state farms were maintained to breed horses for the army, while others have concentrated on producing potable alcohol or on farming the valuable Karakul (Astrakhan) sheep, but they have been especially useful in providing fresh food for arctic mining and other settlements, using forcing-frames and hot-houses and sited to take advantage of local micro-climatological conditions. At Igarka, for example, on the lower Yenisey, a farm lies on a sandy permafrost-free island in the middle of the seven-kilometre-wide river, surrounded by masses of water several degrees warmer than prevailing air temperatures. 'Agricultural towns' (*agrogoroda*) have been built in the Kazakh grainlands for vast new state farms, with workers commuting daily to their work in the fields. The development reflects the deep belief of the Soviet authorities in the ideological as well as amenity advantage of town life over the long-despised village. It was clearly hoped such settlements would encourage people to move into the 'colonial' lands from established communities in European Russia, but it also seems to underlie a change in heart towards the countryside by providing good living conditions in the hope of holding able people on the farm or even of attracting able people from towns into agriculture.

Regional variations in farming types, despite efforts to diversify the pattern and to overcome adverse physical conditions, represent an adjustment to prevailing natural conditions (Figure 5.2). Some differences may also still be attributed to the various farming traditions among the peoples of the Soviet Union. Mixed farming has spread at the expense of other types, expectedly predominating across the main settled belt, though cropping patterns change as one passes from the severe northern winters and poor podzol to the

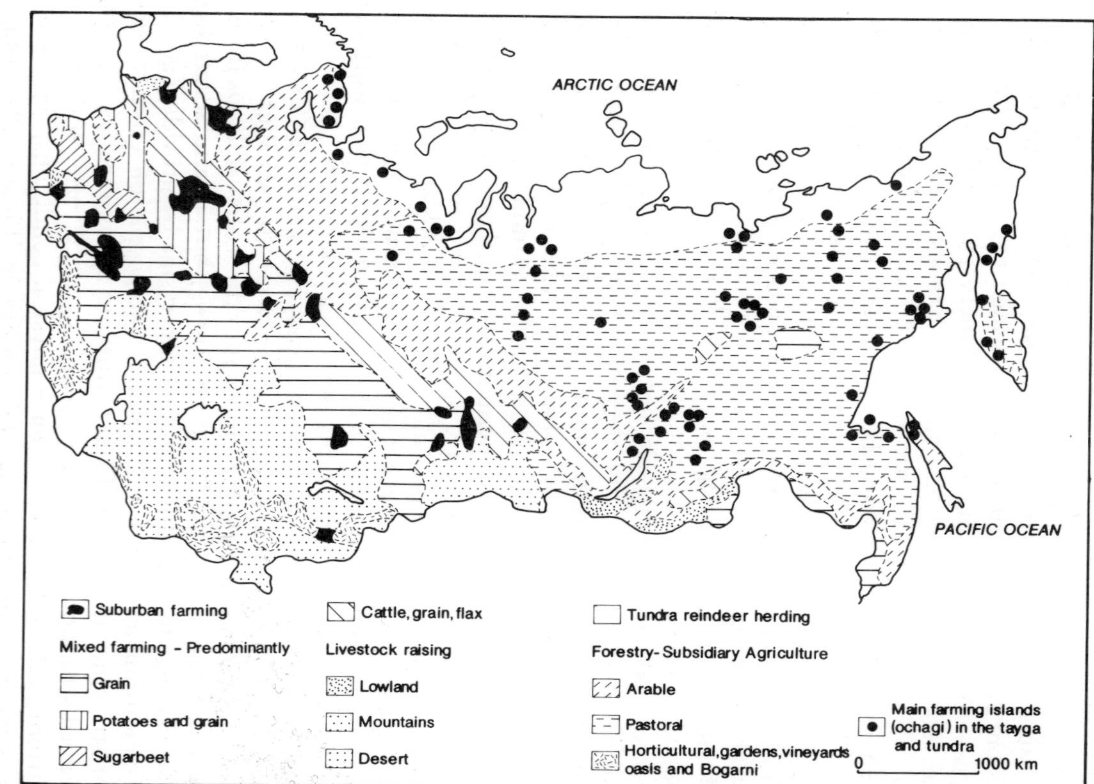

The major agricultural areas show a remarkable coincidence with the pattern of population distribution.

Figure 5.2 *Types of farming in the USSR*

milder winters and warmer summers of the southern black-earth lands, while east of the Volga farming reflects the rising continentality. On lands where mixed farming is not possible, as in the steppe and semi-desert, in the mountains or poorer parts of the tayga, then livestock-keeping is characteristic. The northern forests have only patches of farmed land amid lumbering, while livestock is kept on meadows by the rivers. In some mountain basins and valleys of the south as well as in desert oases, there is garden-type cultivation (*bakhchi*) and near all big towns is a zone of 'suburban' farming (mostly dairying and market gardening) supplying their immediate needs.

The mixed farming belt typifies the principal settled areas of European Russia and Siberia. The northern limit lies in the southern fringe of the coniferous forests, roughly coincident with the northern limit of wheat-growing, a line that has been consistently pushed further northwards. The southern limit is governed principally by increasing aridity, though altitude is also significant in the mountains of Caucasia and southern Siberia. Isolated patches are found in Transbaykalia and in the Far East as well as in the peculiarly favourable soil and climatic conditions of the middle Lena basin around Yakutsk. The north-western part, the Baltic republics and Byelorussia as well as around Leningrad is characterised by flax-growing and dairying, and although summer ripening of wheat is difficult, its area has been expanding. Flax is a demanding crop associated with fodder grasses and dairying (particularly important in the Baltic republics). Rye is everywhere a popular crop since it stands the damp climate better than wheat, and oats are another common crop. Potatoes are also widely cultivated, particularly near Leningrad, where they support a thriving pig industry. Green maize for silage has become significant since the 1950s in a bid to increase livestock holdings. This is generally a countryside where forest and even marsh and rank meadows are interspersed among the fields.

The central part of European Russia grows potatoes, spring wheat and other northerly grains. South and east of Moscow, acid podzol gives way to brown woodland soil and further south to rich black earth, while warmer, drier summers and milder winters combine with the more fertile soils in the wooded steppe to allow a wider range of crops, so that towards the south winter rye, spring and winter wheat, potatoes and sugar beet form the main elements of the crop pattern, the sunflower begins to appear and maize has become more common. Livestock grazes on the meadows by the sluggish rivers and is also fed on silage and waste from agricultural industries, as well as on fodder from complex rotations developed since 1917. This is one of the most intensively farmed parts of the Soviet Union, but the rural overpopulation of tsarist times has been relieved. Even so, the impact of former bad husbandry and overcropping remains in the intensive gullying, but much eroded land has been reclaimed and further loss prevented in this open countryside of fields and occasional forest clumps. Nowadays industrial crops and a greater intensity of livestock farming ease the former overdependence on cereals. Tobacco, makhorka (a kind of tobacco) and hops (near Bryansk) are grown and Michurinsk is a well-known district for apples and pears. The swampy Oka lowlands produce soft fruits and vegetables, while patches of rich, humid soils, as around Vladimir, have extensive horticulture. Throughout this belt, as elsewhere in Russia, bee-keeping (notably in the Mari district) and poultry-raising are significant secondary sources of income.

In Moldavia and the western Ukraine the black and brown earths of the wooded steppe are also mostly mixed-farming country, with an emphasis in many districts on sugar beet, suited to the long warm summers and an annual precipitation of 500–600mm. Winter wheat, grasses and coarse grains are included in the rotations, while waste from the sugar refineries fed to cattle and pigs has resulted in the growth of a meat and dairying industry. The southern part of the mixed-farming belt extends across the Ukraine and the Volga into Siberia, associated with the best black earths and with chestnut and brown soils, to form the main Soviet granary, though conditions and consequently harvests fluctuate considerably from year to year. This belt expanded rapidly late last century as railways were built, and after the Revolution there were several ploughing campaigns to take in extra ground, culminating in the massive Virgin Lands campaign in the 1950s.

West of the Don, winter wheat predominates, as well as sunflower, maize and industrial crops, but rotations now all include grasses. Where too dry for wheat and maize, millet is grown. Despite a lack of natural pasture, waste grain, rotation grass and silage provide for stall-fed animals. In the moist flood plains of the Dnepr, Dnestr and southern Bug some rice is cultivated, and in the sunny sheltered valleys of Moldavia and the Ukraine the vine is grown, though cotton-growing, once quite widespread, has declined. Fruit-growing is notable in Moldavia. The grassland in this southern belt is the natural home for grain, but seasonal problems of moisture and wind already described make cultivation of a good crop hazardous and the success of farming rests on the extensive use of machinery and an assured water supply, both of which have been main aims of agricultural policy in the post-war period. It is a barren, treeless countryside of great fields on the interfluves and the villages hidden in the valleys and gullies.

East of the Don, precipitation is even less in amount and reliability, so that water from wells and irrigation systems is more significant. Nevertheless, the black and chestnut soils of this dry steppe are extremely fertile when correctly watered and cultivated. Spring wheat is preferred because of the danger of frost when the light snow cover is blown away. This is a major supply region for wheat, maize and sunflower, though oil seeds are also significant. Rice is grown in flooded backwaters of rivers like the Kuban, Terek and Don. Along the mountain foot are extensive vineyards. Where there is some grazing, beef cattle, sheep and poultry are kept, and animals are brought in from the Caucasian mountains and the poor steppe for fattening. The Stavropol and Kuban districts saw an early development of large state farms around old Cossack settlements.

On the lower Volga and southern Ural grain-growing and animal-raising are in most respects similar to Northern Caucasia and the lower Don, though wool sheep become important and millet is widely grown in the driest places. On the river below Volgograd, fruit of many types is grown on islands in the braided course. Considerable areas here have been brought into cultivation since the early 1950s, growing spring wheat, millet, maize and sunflower. This same pattern typifies the Virgin Lands scheme, extending far across northern Kazakhstan and southern Siberia to the rich grainlands of the Altay foot.

In Western Siberia, along the railway zone, grain-growing gives way to dairying as the emphasis in mixed farming. Excellent natural pastures exist in the wide valleys of the rivers, the basis of a dairy industry developed late last century, concentrating on articles like butter, easily transported to European markets. This is still pioneer country, where the key to development lies in ameliorative measures to prevent spring flooding and summer drought as well as better transport. In the last forty years mixed farming has spread eastwards into the Altay foothills and the Minusinsk steppe and towards Lake Baykal. East of Lake Baykal, and in the Far East, Russian colonists have abandoned traditional systems to farm Chinese and Manchurian crops with appropriate local methods. Grain is the main crop in Baykalia, but in the Amur-Ussuri basin the pattern is more varied, with crops like soya beans, perilla, millet and rice (mostly around Lake Khanka). In the Zeya and Bureya valleys, sugar beet, cereals and sunflower are grown and pigs and cattle kept, and everywhere there are Manchurian varieties of fruit. Although patches of cultivation have been pushed northwards, a notable cultivated area lies in the anomalously mild climate and good soils of the middle Lena basin around Yakutsk, where there is grain-growing and much cattle- and horse-raising on the excellent meadows.

A major element in the Soviet agricultural scene is pastoralism where arable farming is impracticable or where the former arises from historical conditions. The collective artel has also been adapted to this form of rural economy, though its introduction was often bitterly resisted. In general, pastoralism has been encroached on by the spread of extensive methods of arable farming. The basic distinction in pastoral types is between that of the southern lowlands and mountains of the arid belt and that of the northern forest and tundra fringe.

The poorest steppe and semi-desert are used for sheep-breeding for wool, meat and tallow, though the density of animals is exceptionally low. Cattle are kept in the fringe of the cultivated lands, while pigs are only found among non-Moslem communities, mostly kept by Russians and Ukrainians. Horses were formerly important, though they are still bred, and camels are raised along the Iranian frontier. In winter in Central Asia animals are driven from the mountains to the lowlands, returning in spring. Considerable summer movement from one grazing area to another takes place among animals remaining in the lowlands. The main contribution of the Soviet period has been to increase the carrying capacity of the country by drilling wells to water pasture and by providing drinking pools, while weather forecasting helps to move stocks to safety from storms or drought, with aircraft to maintain contact between the farm bases and the herdsmen. Shelter belts have been planted to provide summer shade and winter shelter. In the Central Asian lowlands the winter pastures lie mostly in the southern valleys, whereas spring and summer pasture is mostly in the Karaganda and Aktyubinsk *oblasti*.

In the Central Asian mountains seasonal movement of herds up and down the slopes remains, but nowadays natural fodder is supplemented by growing suitable crops in the foothill and piedmont zone. Some animals remain in high sheltered basins and valleys throughout the winter, to be joined in summer by flocks from the lower slopes where the grass has withered. Horses are usually kept on the lower slopes, cattle on the middle pastures and sheep on the higher grazings. Caucasian pastoralism is more diverse and greater areas in the mountains are used for crops, while with more forest shelter the density of stock is also higher. Large herds leave the Great and Little Caucasus ranges and the Armenian plateau for winter grazings in the steppe lowlands, to return in late spring, and there are similar movements from the limestone uplands of the Crimea.

Pastoralism in southern Siberia has become increasingly a sedentary occupation as more fodder crops have been grown, with sheep and cattle as the principal livestock, though in the poorer parts of the Altay and Sayan ranges, goats, reindeer and even maral deer are herded. Horse-breeding has remained reasonably important, while yaks are kept along the Mongolian border. Sheep are particularly significant in the Tuva Autonomous Oblast.

In Northern Siberia the old native reindeer-herding remains, but collectivisation and modern methods have been introduced. In winter shelter and fodder are found in the wooded tundra and forest edge, but in summer the herds move northwards, even into the better tundra itself. Modern methods aim to prevent overgrazing the slowly rejuvenating pastures, though these are so poor that several hundred hectares are needed for each animal. Aircraft are used to aid the herdsmen, carrying fodder if required. Planned grazing with short intermediate movements has replaced the old-fashioned exhausting treks.

The vast forests of the tayga elsewhere in Siberia have been nibbled into at the edges for arable farming, while the rich meadows along the rivers are used for livestock. In the forest, though hunting continues, fur-breeding farms have been established in the south. Farming is mostly in Russian or Ukrainian hands, while the native peoples are hunters or herders. This belt contains the bulk of the third of the area of the Soviet Union that is forest (overwhelmingly coniferous) and here lies anything up to one-third of the world's timber reserves. Because of its inaccessibility and its virgin nature, however, immense tracts have little commercial potential. In Western Siberia large areas of forest are interspersed by bog, and in Eastern Siberia and the Far East as much as two-thirds of the timber is overripe. Three-quarters of the timber produced comes from the better stands of northern European Russia, where the annual cut exceeds annual growth, with serious long-term implications. In Siberia, in contrast, annual cut is less than a tenth of annual growth. Half the timber taken is used for building purposes, while large quantities are still used as firewood. The paper and wood-chemicals industries use about a quarter of the annual cut, chiefly in the north-west of European Russia, the Ural, the Baltic and the Far East. The size of the Soviet timber cut each year has considerable impact in the world market, especially for competitors like the Scandinavian countries.

There are some limited areas of special farming types. In Caucasia, for example, growing of crops like tea, citrus fruits, the vine and even tung nuts (for oils) occurs on both irrigated and non-irrigated land. The lower slopes of the Crimean mountains are also renowned for fruit gardens and vineyards (producing some of the best Soviet wines). Irrigation is a distinctive feature of farming in Central Asia, where oases lie in sheltered valleys or along the desert rivers where it is possible to raise two crops a year. Wherever precipitation reaches modest levels, dry farming (*bogarni*) is practised. The emphasis in the Soviet period has shifted to industrial crops, like cotton, which cannot be grown so well elsewhere in the country. There is also rice-growing, and during the interwar years, *kok-sagyz* and related species were cultivated to provide a substitute for natural latex rubber. Few animals other than poultry are kept in the oases, but in Transcaucasia mulberry for silkworms is grown.

The pattern of farming plays a significant role in the long-standing Soviet problem of food supply and the attempts to reduce transport to the minimum. The Soviet system does not yet seem to have solved the problem of quick and effective distribution, so the problem of moving perishables is a critical one. For this reason the suburban farming zone immediately around all reasonably sized towns reflects the area from which they draw much of their perishable foodstuffs like milk and meat, fresh vegetable or fruit. From this belt and immediately beyond food is also carried into the collective farmer's 'free markets'. The whole distribution and supply problem is exacerbated by inadequate food-processing on a modern scale, for the range of tinned or deep-frozen foodstuffs is still markedly behind Western levels. It is perhaps fortunate that people in the Soviet Union continue to enjoy traditional salted, dried or pickled foods and such items as sour cream (*smetana*), all of which are easily stored and are transportable without elaborate facilities. In the arctic and sub-arctic, as already noted, expensive farms have been developed to supply local workers with fresh food, but this helps by reducing the need to transport foodstuffs into these remote areas, a task which would be made doubly difficult by the relatively low level of food-processing elsewhere in the Soviet Union. Regional food supply also has a broader dimension inasmuch as natural conditions make it impossible to produce the same array of foodstuffs in every part of the country, and consequently regional interchange of foodstuffs cannot be completely eliminated, with long-standing areas of surplus and deficit. A typical example is the shipment of large quantities of grain from the steppe and wooded steppe northwards into the forest belt, less suited to its cultivation and yet with large urban populations requiring bread grains. Another example, drawn from a rather different but cognate field, is the shipment of fish, mostly in processed form, deep inland from the fishing ports.

Certainly, extension of the cultivated area in Siberia and the Far East has been closely related to the need to provide food locally for the growing urban populations as these regions have developed their mining and industrial potential. Western Siberia in particular has turned from a net exporter of foodstuffs to a net importer. Several vital non-food crops have been concentrated in areas best able to produce them, even at the expense of local cultivation of food crops. Cotton, once far more widely grown in southern European Russia, is now overwhelmingly concentrated in Central Asia. Some students have suggested this is done not only for more efficient production, but also because it has consequently made this potentially politically unreliable area more closely dependent on the rest of the Soviet Union for food supplies. Other regional specialisation is reflected by the production of about 60 per cent of refined sugar from the western part of the Ukraine, and together the Ukraine, Moldavia and Northern Caucasia produce over half the canned-food output of the Soviet Union.

To judge from considerable Soviet imports of grain and other foodstuffs from both Comecon and the Western world, despite all the effort put into agriculture, the sheer physical adversity of the natural environment and other constraints leave it as the Achilles heel of the Soviet economy.

5.1 Where to follow up this chapter

Extensive discussions can be found in Laird, R. D., *Soviet Agriculture and Peasant Affairs* (University of Kansas Press, Lawrence, 1963), McCauley, M., *Khrushchev and the Development of Soviet Agriculture – The Virgin Lands Programme 1953–1964* (Macmillan, London, 1976), Millar, J. R. (ed.), *The Soviet Rural Community – A Symposium* (University of Illinois Press, Chicago, 1971), Symons, L., *Russian Agriculture: A Geographic Survey* (Bell, London, 1972), and Volin, L., *A Century of Russian Agriculture – from Alexander II to Khrushchev* (Harvard University Press, Cambridge, Mass., 1970).

A Soviet view is expressed in Bogush, G. M. and Shaykin, B. G., *Selskoye Khozyaystvo SSSR* (Moscow, 1977), which contrasts with the view gained from *USSR Agriculture Atlas* (Central Intelligence Agency, Washington, D.C., 1974). Also useful is Rakitnikov, A. N., *Geografiya Selskogo Khozyaystva* (Moscow, 1970).

6

Soviet Industry – The Resource Base

The shortcomings and problems of Soviet industrial development have been eased by the remarkably fortunate resource base compared with some of the other major industrial nations. Although the Soviet authorities are reluctant to publish comprehensive information about their reserves and production of most minerals, when the scanty available data are pieced together the result is impressive, to which must also be added a considerable endowment among other members of Comecon. There are certainly some raw-material deficiencies, made more difficult because the materials in question are commonly substitutes for each other. Even so, the richness and diversity have been sufficient to give Soviet planners plentiful encouragement to contemplate a virtually closed economy and a high level of self-sufficiency. This aim is certainly realistic if some of the considerations of cost effectiveness in resource development applied in the capitalist world were disregarded. The over-all national totals nevertheless conceal several difficulties: many deposits are of relatively low grade or have problems of working because of their geological character; some promising deposits are located in remote places with unusually harsh physical conditions (Figure 6.1). Mere presence of a particular mineral deposit is not necessarily an indication that it will either be workable or even be worked; a major decision has often to be made as to whether a mine should refine or process its crude in whole or part on the site or whether the crude ore should be shipped away to some industrial district where more sophisticated methods can be employed or where energy requirements for processing are more easily satisfied. This has important implications for the more

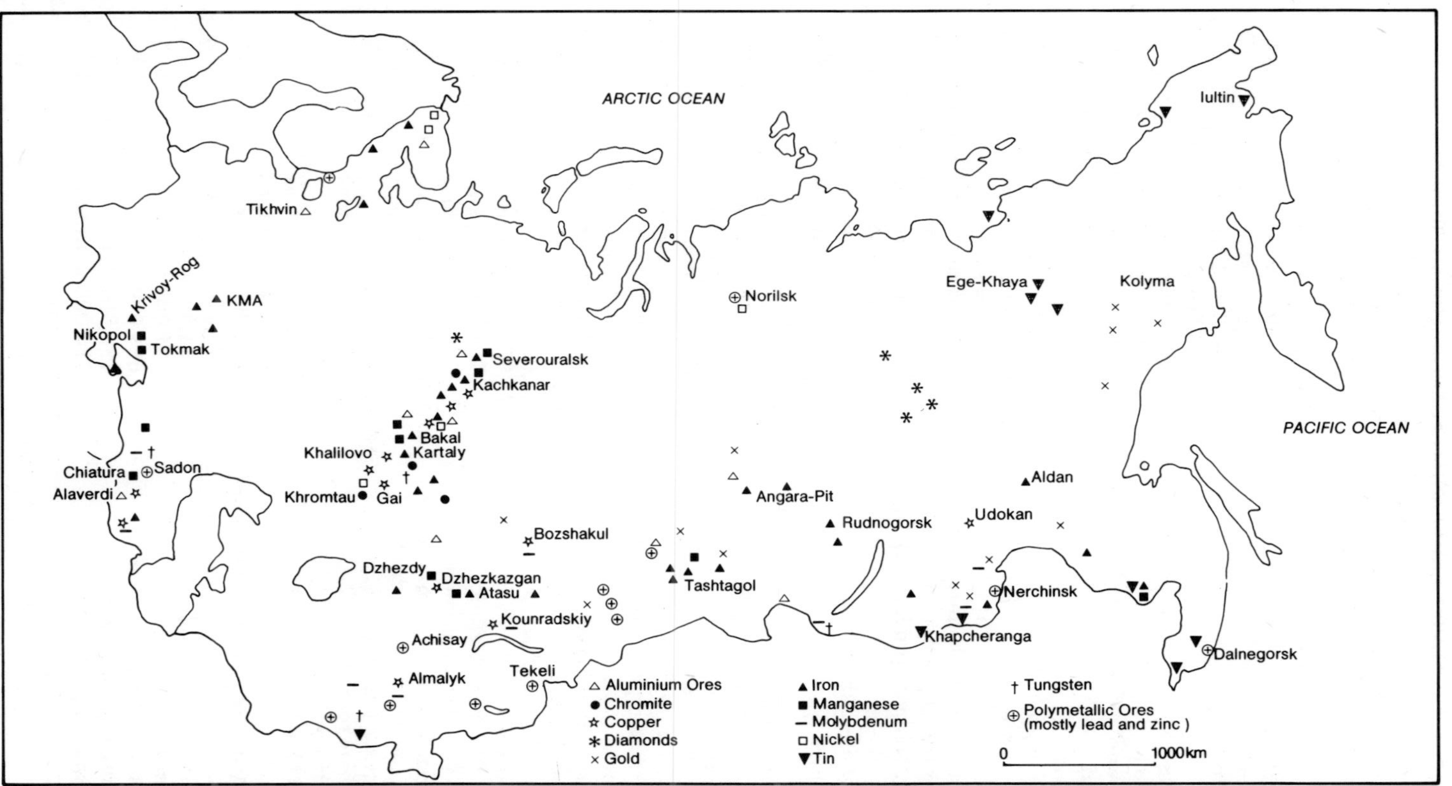

The map reflects both the variety and wide distribution of metallic mineral deposits in the Soviet Union.

Figure 6.1 *Mineral deposits in the USSR*

Table 6.1 *Soviet mineral production and its world share*

	Output	World share
Aluminium	1,760,000 short tons	13%
Antimony	8,500 short tons	11%
Asbestos	2,520,000 short tons	45%
Barite	440,000 short tons	8%
Bauxite	4,400,000 short tons	6%
Alumina	2,800,000 short tons	9%
Cadmium	3,000 short tons	16%
Cement	136,651,000 short tons	17%
Chromite	2,337,000 short tons	25%
Coal – bituminous	511,500,000 short tons	25%
Coal – lignite	190,000,000 short tons	19%
Cobalt	1,950 short tons	9%
Copper (mine production)	880,000 short tons	11%
Copper (smelter production)	880,000 short tons	11%
Copper (refinery production)	837,000 short tons	10%
Diamond – gem	2,000,000 carats	19%
Diamond – industrial	7,900,000 carats	27%
Diatomite	460,000 short tons	24%
Gold	7,700,000 troy oz.	20%
Iron ore – gross weight	235,225,000 long tons	27%
Iron ore – metal content	138,783,000 long tons	27%
Iron – pig	115,086,000 short tons	21%
Iron – raw steel	159,620,000 short tons	21%
Lead (mine production)	520,000 short tons	14%
Lead (smelter production)	550,000 short tons	15%
Manganese (35% Mn)	9,370,000 short tons	34%
Mercury (76-lb flasks)	56,000 short tons	23%
Mica (all grades)	95,000,000 lb	20%
Molybdenum (contained Mo)	20,600,000 lb	11%
Natural gas	11,950,000 million cu. ft	20%
Nickel (mine production)	154,000 short tons	17%
Nickel (smelter production)	176,000 short tons	21%
Nigrogen fertiliser – consumption	8,110,000 short tons	17%
Nigrogen fertiliser – production	9,331,000 short tons	19%
Oil – crude (42-gallon barrels)	3,822,000,000	18%
Phosphate rock (34.8% P_2O_5)	26,700,000 short tons	23%
Platinum group	2,800,000 troy oz.	47%
Potash (K_2O equivalent)	9,150,000 short tons	34%
Salt	15,432,000 short tons	8%
Silver (smelter/refinery production)	44,000,000 troy oz.	14%
Sulphur	8,270,000 long tons	17%
Tin (mine production)	31,000 tonnes	14%
Tin (smelter production)	31,000 tonnes	14%
Tungsten (mine production)	17,600,000 lb	19%
Zinc (mine production)	790,000 short tons	8%
Zinc (smelter production)	790,000 short tons	13%

Source: *Minerals Yearbook, Vol. I* (US Department of the Interior, Washington, D.C., 1976).

remote mines in the harsher parts of the Soviet Union, while the economic implications of weight loss in mineral processing are particularly significant in the transport and distance problems of the Soviet economy.

6.1 **Energy resources** (see Figure 6.2)

One of the most promising features from the Soviet point of view is the well-endowed potential of energy resources. The key problem here is that about 87 per cent of all energy resources lie in the eastern regions, whereas the proportion of utilisation is overwhelmingly in the west; the implications for transport are obvious. Of the fossil fuels, the Soviet Union has immense reserves of coal of several types and adequate for many centuries ahead. Petroleum resources have also been substantially augmented, though many are in difficult and remote physical environments. There is some doubt as to whether further discoveries will keep pace with anticipated needs, though certainly the Soviet Union will remain one of the world's largest producers well into the next century. Reserves of natural gas, separately or in association with oil deposits, are extremely large, though again some are in remote locations. So far as can be judged from scanty information, the Soviet Union is also well endowed with materials for nuclear fuels. One of the greatest potential energy resources is in the immense and largely untapped reserves of water power, though here location and environmental conditions comprise a serious long-term constraint. Of lesser and usually only local significance are deposits of oil shale and peat, while even firewood plays a minor but not inconsequential role. Regional development has been greatly influenced by the geographical distribution of these energy resources, and in many places, in order to establish some sort of local self-sufficiency, small and unsuitable deposits have been worked at high cost. One of the most serious problems in energy supply is that through the differences in the spatial distribution of energy resources and energy consumers, difficulties in transport and supply commonly arise. In both the Ural region and the central industrial region around Moscow, inadequate local energy resources have been a restraining factor for industry, especially for large-scale iron and steel making in the former region. There is also the growing concentration of investment in highly energy-intensive industries like electro-metallurgy and electro-chemicals in Siberia to use its immense energy potential. The major role assigned to energy production, even from unpromising sources, in regional planning strategies has already been referred to in Chapter 2, though in spite of such a policy a

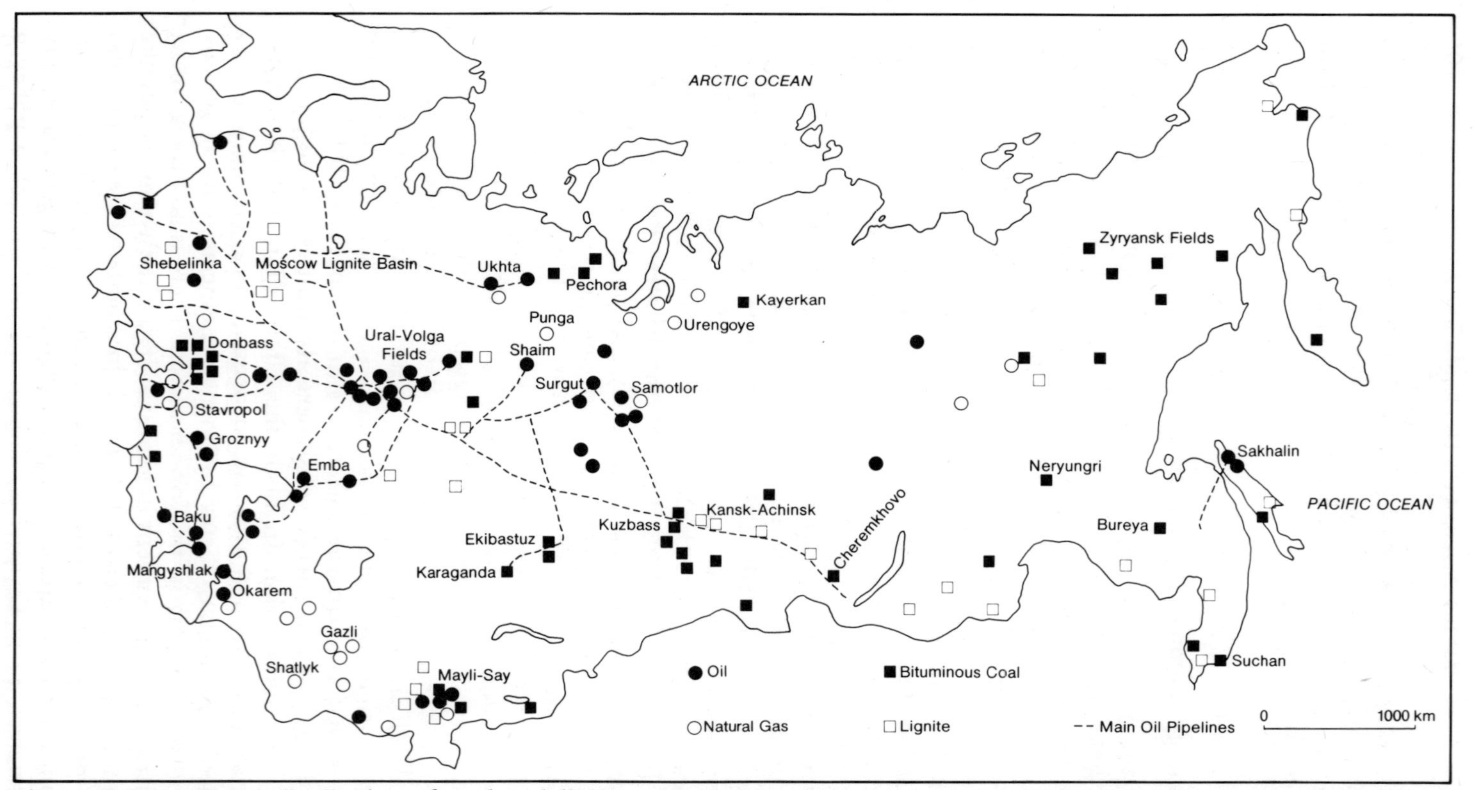

The wide geographical distribution of coal and lignite mining is now being matched by that of oil and natural gas, the result of major developments since the 1960s.

Sources: various Soviet atlases and other maps.

Figure 6.2 *Energy resources in the USSR*

relatively small number of the major economic planning regions still provide the bulk of the total national energy supply.

Production of all fuels expressed in standard physical units has increased just over thirty-four-fold since the Revolution. Since the inter-war years petroleum and natural gas have shown a proportionately greater growth than coal, the mainstay of energy generation until the 1950s, when its share shortly thereafter began to decline. By 1970 it had been overtaken by petroleum and by the early 1980s it will similarly have been surpassed by natural gas. By the end of the 1970s, however, the role of coal was being reassessed to try to find it a more significant future contribution in the energy balance. Well over half the coal consumed is used as an under-boiler fuel and its decline has been associated notably with its replacement in railway transport and in electricity generation. Before the Revolution a large part of the petroleum produced was exported and its real share of the internal energy balance was probably below 10 per cent. The relatively late growth in the role of petroleum and natural gas has several underlying causes, of which one of the most important is that it was possibly not such an apparently cheap source per unit of energy produced under Soviet accounting and costs systems compared with those of the Western world. Certainly the structure of Soviet transport did not create such a strong demand as in the West and the whole Soviet chemicals industry, including petrochemicals, remained markedly backward well into the 1960s, just as the Soviet oil and gas industry in both exploration and exploitation as well as refining lagged technologically behind the West.

6.2 Spatial distribution of the main energy resources

Coal

Total Soviet reserves of all types of coal are estimated to be sufficient to last for over 1,200 years at the present annual rate of production; if only the highest-rated reserves in terms of quality and accessibility are included, then they are adequate for at least the next 250 years. Figures of reserves are far from the whole story: the size of the individual deposits and their characteristics as well as location are notably significant. The striking feature of any map of Soviet coal resources is their wide geographical scatter, with few parts completely devoid of deposits. Nevertheless, almost 94 per cent of total geological reserves (64 per cent if measured in proven reserves) lie east of the Ural ranges, whereas over 50 per cent of total annual production comes from the smaller though formidable reserves in European Russia. The emphasis in regional planning on maximising energy generation from local resources and making the lowest possible demand on transport has resulted in many small deposits being worked whose quality is often quite poor and production costs high, especially as mining methods are usually backward.

The coal supply of European Russia is dominated by the Donbass field, an area of some 60,000km^2, whose limits have been extended westwards as well as eastwards with more detailed prospecting. This is the largest coal-mining district in the Soviet Union and contains about a quarter of the total explored reserves. Donbass coal is of high quality, with about 30 per cent of the reserves in the basin of anthracite and some 25 per cent of coking coal, and it produces about 60 per cent of all Soviet metallurgical coke. The mines are generally large and shallow, with relatively straightforward conditions apart from dust and gas. With good railway links, the coal has been widely distributed throughout European Russia, while the field itself has attracted much heavy industry and its thermal power stations are notably important.

In the far north-east of European Russia the Pechora coalfield has grown in importance since completion of the railway to Vorkuta in 1942. The coal is also high quality, if not quite as rich as the Donbass. Although its total reserves are greater than the Donbass, its economically exploitable reserves are more modest, but output per mine is higher, despite difficult environmental conditions. It is particularly significant because of its large proportion of coking coal, otherwise becoming difficult to obtain. Over 70 per cent of workable reserves lie around the main mining towns of Vorkuta and Inta and some working is by opencast methods. Pechora coal is becoming increasingly important in north-west European Russia.

West and south of Moscow, a shallow basin of brown coal of low calorific value and high in ash has been worked, mostly for electricity generation because of its proximity to Moscow, but mining is declining as costs are higher than for natural gas piped from other regions. Since 1945 mining has begun on a bituminous coalfield near Lvov on the Polish frontier whose output is consumed mostly in the western Ukraine and in Byelorussia. Elsewhere in European Russia, coal, rich in ash, occurs at Tkibuli and Tkvarcheli in Georgia, playing a significant role in Transcaucasian heavy industry, but reserves are limited, while Akhaltsikhe brown coal is used for electricity generation. Although the Ural region is richly endowed with metallic ores, it is poorly provided with coal in small occurrences on either flank of the ranges. Around Kizel near Perm, coal used for electricity generation lies deeply buried and mining is arduous. Brown coal east of Chelyabinsk is also useful for power generation, while extremely wet deposits in the south are worked by the opencast method.

In the eastern regions coal-mining is dominated by the Kuzbass field of Western Siberia, covering some 70,000km^2. Thick seams of high-quality coal, some of outstanding coking type, make this one of the finest fields in the world, and there is considerable potential for the future. The coal is sent by rail widely throughout Siberia, but it moves primarily westwards to the heavy industry of the Ural, while most recently it has been used to replace declining Donbass supplies in European Russia. Not far east lie the vast brown coal deposits of the Kansk-Achinsk field, as yet not fully explored, with mining around Itat, Bogotol, Nazarovo and Borodino. Low in ash and of reasonable calorific value, these brown coals have great importance for electricity generation, especially as this is one of the cheapest producers in the USSR. The gas coal of the Minusinsk Depression is also used for power stations, while several fields occur west of Irkutsk, of which Cheremkhovo is the main producer. Unfortunately, these coals have much sulphur, but are reasonably cost-effective through the use of either opencast or shaft mining.

Several modest deposits of varying quality lie in Transbaykalia, but the largest future potential in all Siberia lies in vast ill-defined northern basins. The Tunguska fields, for example, lie within an area of one million km^2, possibly containing up to 1,745,000 million tonnes of coal. Remote and difficult of access, these deposits are still inadequately known and scarcely exploited. Deposits in the Taymyr Peninsula are even less known, though they are mined near Norilsk. Another large basin lies along the Lena, probably containing coal mostly suitable for power stations, but little explored and almost completely unworked. In southern Yakutia, in the Aldan basin, reserves of 40,000 million tonnes have been assessed, of high quality, some suitable for coking. With the building of a branchline from the Trans-Siberian railway, these deposits have begun to be worked by opencast methods using Japanese equipment at Neryungri near Chulman, and the coal will go to Japan. The Bureya field, whose output already goes for industrial and power station use in the Far East, augments coal from the Suchan mines near Vladivostok. Brown coal is also mined here in the Uglovaya field. Variable and scattered deposits in Sakhalin are locally significant, like small workings in the far north-east of Siberia.

Over 500 separate deposits of bituminous and brown coal are known in Kazakhstan and Central Asia, but the main producer is Karaganda, where thick seams of coking coal near the surface can be worked by opencast methods. The third major coalfield in the Soviet Union, its output goes principally to the Ural or to local industry. Recently, good deposits at Ekibastuz near Pavlodar have been worked cheaply by opencast methods, while a large potential is seen for the Maikuben brown coal deposits not far away. Kushmurun near Kustanay has brown coal and semi-bituminous deposits, again worked in opencast pits with seams 80—100m thick. Varied coals in scattered deposits lie elsewhere in Central Asia, economically most important in the Fergana valley. The Uzgen field contains good coking coal, but it lies in difficult mountain terrain. Angren in the Uzbek SSR serves Tashkent, and the brown coal here is worked inexpensively by opencast methods. The Tadzhik SSR has several small deposits that include some coking coal.

Oil

With an inflow of foreign capital late last century, oil production rose quickly, and by 1903 Russia was the world's largest producer, overwhelmingly from the Caucasus. Recovery after the Revolution was slow so that before 1939 the Soviet Union had not regained its former position, but growth since 1945 has once again put it at the top of the producers' league and it has become a significant exporter of crude and refined products. Although the real extent of the oil reserves is uncertain, it looks as though they are sufficient for about sixty years at present production levels. Since the late 1950s the rate of growth of crude-oil production has grown faster than proven reserves, and although this does not imply any impending shortage, there are some problems of equating supply and demand. Certainly, even with a steady growth in crude output, it seems unlikely that the Soviet Union will find it easy to satisfy growing home demand and still maintain exports at the level of the late 1970s. Some observers even believe the Soviet Union will have to become a *net* importer and compete for supplies with Western countries in the Middle East and Iran. About half Soviet territory has the geological characteristics in which oil and gas fields could occur and there are large areas of continental shelf in the adjoining seas so far completely unexplored, so discovery of further significant deposits therefore does seem likely.

Until the mid-1950s known oil resources and the working of deposits lay overwhelmingly concentrated in the western part of the country. Until the inter-war years dependence was almost entirely on the deposits of the Caucasian isthmus (notably around Baku), but by the early 1940s large discoveries were being made in the Volga basin and along the western foot of the Ural. So great was the potential here that the fields were for long known collectively as 'Second Baku', so that their comparability with the main producer up to that time might be understood. From the mid-1950s interest shifted to exciting possibilities in inhospitable forests and swamplands in the west Siberian lowland. Deposits elsewhere had generally only local significance, though even so the strategic undertones of resources in Sakhalin, Central Asia or the Arctic were substantial. As the geography of the major oilfields has shifted, so has the locational pattern of refineries. Until the mid-1950s the bulk of refining capacity lay in the Caucasian region, but through the 1960s a major refining and petrochemicals artery appeared along the Volga, and now refining is becoming significant in southern Siberia. The working of deposits in the more inaccessible areas has been eased by building pipelines to the main refining and consuming districts.

The oilfields of the Caucasian isthmus, mostly in Tertiary or Cretaceous strata, date back to the 1860s, notably around Baku, which remains a principal centre, whose wells lie in the Kura lowlands and the anticlinal structures of the Apsheron Peninsula and their offshore continuation. Offshore wells in the Caspian Sea are becoming more important and their development is using much Western technology, particularly as they move into deeper water. Similar strata and structures mark the northern flanks of the Great Caucasus, where the Kuban field on the western side began as early as the 1860s, but the more easterly fields (e.g. Groznyy) came around the 1890s. Most recently, the Stavropol area has developed as one of the largest natural gas fields in the USSR.

In the Volga–Ural oilfields deposits occur in relatively shallow old sediments overlying the crystalline platform. It was not until the inter-war years that extensive exploration began, and the German wartime threat to the Caucasian fields gave a major impetus to development. The oilfields, mostly in anticlinal structures, lie conveniently in a triangle between Moscow and the Donbass on the west and the Ural on the east, well served by rail and water transport. By the early 1960s these fields were producing over 70 per cent of all Soviet crude oil, but this has fallen to something under 50 per cent (nevertheless substantial compared with the tenth from the Caucasian fields). Most of the crude is refined locally, though shipments are made via pipeline to Eastern Europe and to Siberia.

The other fields of European Russia are either relatively small or so recently discovered that little development has yet taken place. New oil- and gas-bearing areas lie between the Dnepr and the Donets in the Ukraine, with discoveries in Byelorussia and the Baltic littoral, while older workings occur along the northern Carpathian footslope. Large gas accumulations have yet to be opened up. In northern European Russia a small field with a considerable local role occurs around Ukhta, mostly developed since the 1940s, and to the south near Troitsko-Pechorsk lie large gas deposits.

Associated with the Caspian depression are two important oilfields in Kazakhstan. Around 1911 oil deposits began to be worked between the lower reaches of the Ural and Emba rivers, with output nowadays mostly piped to the Ural industrial towns or for processing at Guryev. Most recently vast reserves have been worked in extremely arid, hostile country on the Mangyshlak Peninsula. Structurally associated with the Apsheron Peninsula but lying on the eastern Caspian shore, significant deposits occur notably in the Cheleken Peninsula and near Krasnovodsk, while Nebit Dag has substantial gas reserves. The prospects for further worthwhile deposits here are considered good. Elsewhere in Central Asia, small oilfields occur in the Fergana basin, in the Amu Darya valley near Bukhara and Khiva and in several depressions in the Tadzhik mountains, with extremely large reserves of gas, notably at Gazli near Bukhara.

The importance of Siberia rose formidably during the 1960s, and over thirty fields are now recorded. The major reserves of oil and gas all occur in the deep, young sediments of the west Siberian lowland. Two main groups of deposits were first identified – one in the middle reaches of the river Ob and the other near Shaim, some 500km north of Tyumen. A third major area began to emerge in the late 1970s in the north-east of the lowlands in the basin of the river Taz. The gas deposits of these lowlands are of particular interest because of their nearness to the industries of both Western Siberia and the Ural region. Well over a quarter of all crude-oil production in the USSR may now be coming from Western Siberia, where a massive pipeline construction programme is under way. The development of these deposits faces serious environmental difficulties, with trying swampy terrain and permafrost, besides the difficulties of accessibility to conventional forms of transport, while the building of the necessary infrastructure for the construction and exploitation camps is outstandingly expensive. Grandiose plans once suggested to turn the swamps into a giant reservoir to supply water to arid Central Asia clash with the exploitation of the oil and gas resources and appear to have been quietly shelved.

An oilfield of considerable potential has been located north of Irkutsk in the Lena basin and is already worked at Markovo for high-quality crude from Cambrian strata. The Vilyuy basin also holds promise of oil, though so far only gas has been located and is piped to Yakutsk. Some small workings have been opened in even more harsh terrain in the Khatanga valley of the Arctic. In the Far East the oilfields of Sakhalin, developed for strategic reasons before 1939, are of growing interest.

The Soviet Union's reserves of natural gas are the greatest in the world and becoming one of the main energy sources for industry, with a growing network of pipelines from the main fields focusing upon the principal industrial districts. A constraint to the expansion of the use of natural gas comes from the resource consumption in constructing additional pipelines, partly because of inadequate capacity to produce suitable pipes. Natural gas has made a particular incursion into thermal power stations, where it has increasingly replaced low-grade brown coal, but it has been questioned whether natural gas should not be used more selectively. There are also resources of oil shales, notably developed on a considerable scale in Estonia, but the Volga deposits have not yet been used intensively.

Electricity

In the Soviet Union electricity has a particular ideological significance arising from Lenin's dictum that 'Communism is Soviet power plus the electrification of the whole country.' The massive increase in output and consumption since the Revolution follows a world-wide trend, but in terms of production per head the USSR remains at the lower end of the spectrum for the advanced industrial countries.

Electrification has been a key element in regional planning, with an early attempt to define economic regions around principal generating regions (see Chapter 2). Supply poses several problems in the spatial dimensions of the Soviet Union: first, there is the decision whether to give preference to hydro-electric or thermal generation; second, there is the technological problem of the creation of regional grids and especially of a national grid in the immense distances of the country; and third, in the case of thermal generation, the decision of what fuels to select. The choice between thermal and hydro generation is related to whether the high capital costs of construction but low running costs of hydro plants are preferable to the lower capital costs of building thermal stations but of accepting the higher unit generating costs, and there is also the load pattern to be served by the stations to be considered.

The construction of massive hydro-electric power stations has been used as powerful propaganda to advertise the achievements of Soviet technology, though in spite of this since about 1958 favour has swung towards thermal power stations which can usually be constructed more quickly. At the end of the 1970s about a fifth of all current was derived from hydro-electric generators, even though the installed capacity represents only a small fraction of the total potential. This resource suffers, however, from the common Soviet dilemma that production and consumption are separated by great distance and formidable environmental barriers. By far the greatest part of the reserve of hydro-electric power potential lies in Siberia,

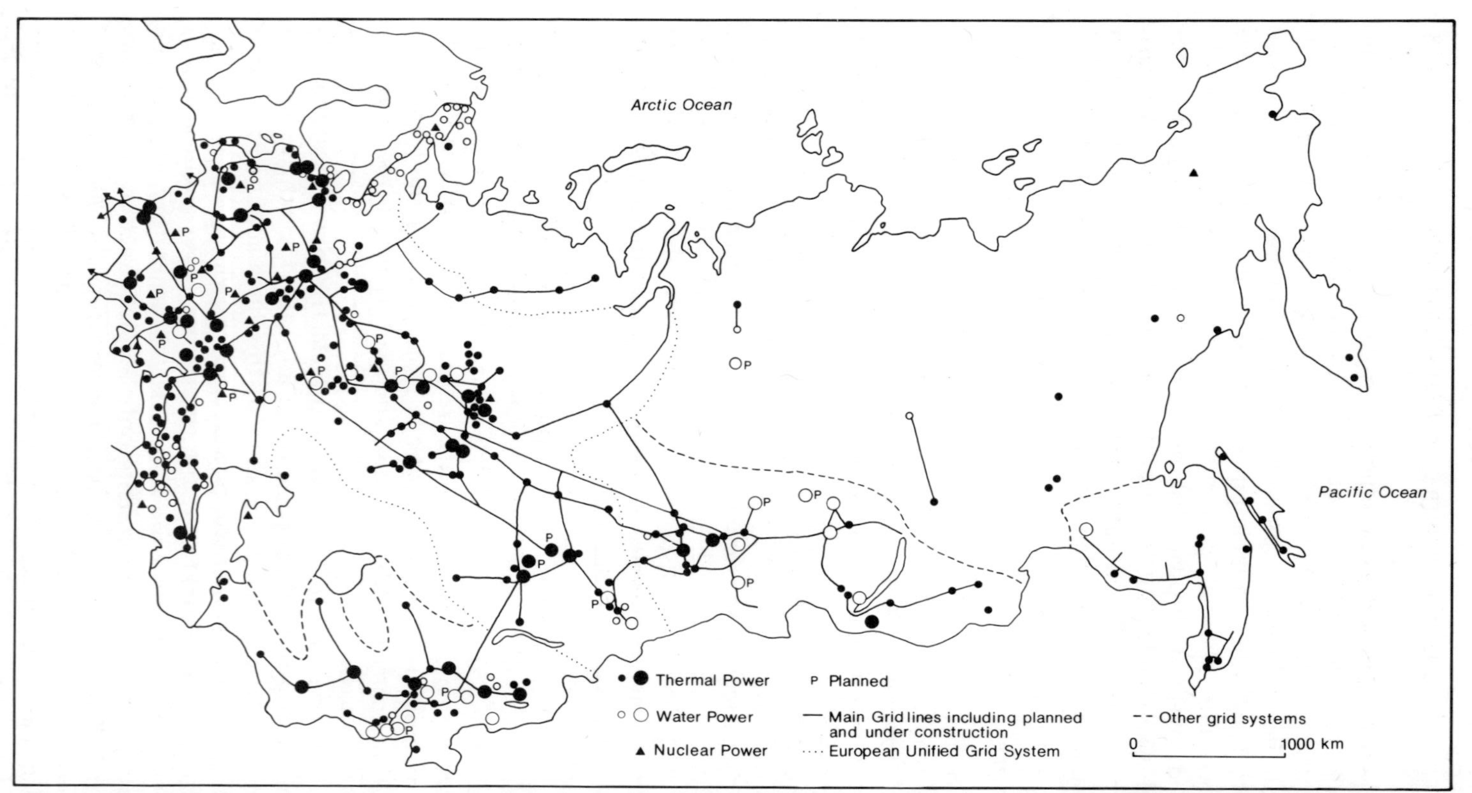

Since the time of Lenin, electricity has occupied a special place in Soviet development. The map reflects the realisation of a nationwide grid system.

Figure 6.3 *Electricity supply in the USSR*

where the severe climatic conditions, particularly the problems of a long, deeply frozen winter, and general remoteness are a deterrent to development, so it is consequently unlikely in the near future that the present tenth of the total resources developed will be greatly increased. The classic pattern of hydro-electric development has been to build 'cascades' of barrages along selected rivers, often in association with navigational improvement or irrigation. In European Russia, where over 70 per cent of all current is consumed, major systems exist on the Dnepr and on the Volga–Kama waterways, but small cascades have been built on the Karelian rivers and there are also hydro-electric barrages on some Baltic rivers. The largest cascade under development is on the Siberian Angara–Yenisey river system, where the large Bratsk plant is one of the most publicised post-war achievements. When complete, this system of barrages will include the largest in the world on the Yenisey near Ablakovo. Groups of hydro-electric barrages have also been built on the upper reaches of both the Ob and Irtysh rivers, but projects well downstream are now unlikely to be started since they could clash with oil exploitation in the west Siberian lowland. Water power gives Siberia its role as one of the world's greatest storehouses of potential energy supplies.

Some particularly useful hydro-electric sites lie in the mountains of Central Asia and Transcaucasia. In the latter, cascades have already been built on the Rioni and Razdan rivers, while one of the largest stations is the Mingechaur barrage on the Kura river. In Central Asia, the Naryn and Vakhsh rivers have significant cascades of moderately sized stations. Several have been designed in association with irrigation, whose requirements for water are given priority, resulting in relatively low utilisation for electricity generation.

Over 80 per cent of all current is generated in thermal power stations using several different sources of fuel and located in every part of the country. The wide scatter of these stations accounts in many instances for the working of small, low-quality coal deposits and they have also been developed to burn low-grade fuel like coal dust in the Donbass, oil residues at Baku and sawmill waste at Arkhangelsk. In Estonia some power stations are fired by oil shale, and east of Moscow peat-fired power stations have been used. The growing use of oil and natural gas, easy to transport, has made them attractive to use in power staions, especially as they have many advantages over low-grade fuels like brown coal. Thermal power stations have tended to become progressively larger, able to give considerable economies of scale and to serve greater areas, but in its turn this has demanded better distribution systems through the building of regional grids. The Soviet Union also claims to have developed the first operational nuclear power station (Obninsk near

Moscow, 1954), but nuclear generators have been otherwise rather slow to develop. Plants have been built, however, at Voronezh and Melekess to boost electricity supplies in the Black Earth and Volga regions and for the Ural region at Byeloyarsk near Sverdlovsk. A special type of plant for remote regions has been in use at Bilibino in north-east Siberia. Several additional plants are now planned for western European Russia, where supply of conventional fuels raises transport demands, as in Lithuania, Byelorussia and the western Ukraine, while other stations are to be built in the Kola Peninsula, near Leningrad, and in Armenia. Some Western observers have suggested that development of nuclear power stations (despite access to raw materials for fuel) has been less critical for the Soviet Union than for many Western industrial nations because of its big reserves of conventional fuels and water power.

With wide time-zone variations across the country, the use of a national grid system to even out peak loadings could help Soviet energy-utilisation patterns. For example, although the daily load pattern in Moscow and Novosibirsk is much the same, the peaks occur about three hours later in Novosibirsk through its more easterly position. If current could be transmitted back and forth between these two major cities, worthwhile savings could be made. Although large regional grids have been built, focused essentially on the main electricity-generating and consuming districts, a national grid depends on the development of appropriately high-voltage transmission, a field in which Soviet technologists have made great progress. A basic 500kV system capable of having its voltage subsequently raised is being built to join the main European centres to those in Siberia. Of the interregional grids already established, the largest are the Central – linking the principal industrial centres like Moscow, Gorkiy and other big towns together – and the Southern, covering the Donbass, Dnepr and Kharkov districts as well as the Volgograd hydro-electric barrage. Another significant grid system centres on Leningrad for the north-west and a powerful grid has been developed to serve the Ural industrial towns, while Transcaucasia is distinguished by its own system. These systems are also being joined to the West Siberian and Kazakh systems, which will be linked to Central Asia, where a grid has already been partially built. An as yet self-contained grid is being built in the southern part of the Far East. The Soviet Union is also a participant in the *Mir* electricity grid linking together the Eastern European members of Comecon.

6.3 Metallic minerals

Although intensive exploration has taken place, large areas still remain to be assessed in detail, and even as late as the 1950s several tracts in north-east Siberia had not even been covered by reconnaissance surveys, while worked deposits in many instances still need to be completely investigated. The Soviet authorities are reluctant to publish detailed information about their resources, so that figures for reserves and output frequently depend on little more than conjecture. Nevertheless, the Soviet Union is unquestionably well endowed with metallic minerals, despite some deficiencies, which in most instances nowadays are conveniently covered by trade within Comecon. Estimates of total reserves, however, do not always reflect the problem that some of the most metallic ores in scarce supply are in remote and environmentally hostile locations.

Large reserves of *iron ore* occur, generally better in quality than most currently worked in the Western world, but some of the biggest resources as yet unworked lie unfortunately in remote locations. *Manganese* and *chrome*, vital alloy metals, are abundantly available and the Soviet Union is a major producer of chrome and an important exporter of manganese. Adequate deposits of *vanadium, titanium* and *nickel* exist. The equally important alloy metals, *cobalt* (mostly associated with nickel in the ratio one part cobalt to seventy parts nickel), *molybdenum* and *tungsten*, appear unfortunately scarce as they act as substitutes for each other in high-speed steels. Molybdenum may be obtained from China and North Korea; a little cobalt may be received from East Germany, while China, a major world producer, could also supply tungsten.

A special Comecon directorate deals with non-ferrous metals, suggesting that the position in the *bloc* is not satisfactory. In the early 1960s *copper* supplies for a time fell well behind growing requirements, especially as railway electrification got under way, but exploration and development of new sources (notably also the building of additional smelter and refinery capacity) has changed the position from a net importer to a net exporter and the USSR may now be the world's second largest producer. *Lead* and *zinc* commonly occur together in complex ores along with other minerals, notably *copper* and *silver*. The reserves and supply position appear satisfactory, though working smaller and remoter deposits to cover rising needs for lead may appreciably raise costs. Minerals for *aluminium* seem generous, and a considerable resource base in Eastern Europe is also accessible, so that aluminium alloys are consequently commonly substituted for scarcer metals. The *tin* position has also improved through the exploitation of remarkably inaccessible deposits in north-east Siberia and even a little is exported. *Platinum* metals, *mercury* and *magnesium* metals all seem to cover Soviet needs. The Soviet Union is nowadays probably one of the world's larger *gold* producers. Far less is known about *silver* but it seems available in sufficient quantities, mostly in association with polymetallic ores.

6.4 Iron and the ferrous metals

Of the iron-ore deposits workable under present conditions, some 60 per cent are in European Russia, but it is the eastern regions that hold the most promise. The greatest worked deposits are at Krivoy Rog in the Ukraine, with most ore going to the Donbass steelworks, though increasing quantities have been sent to Poland and Czechoslovakia. A vast potential, yet to be fully assessed, lies a little further north around Kursk, Belgorod, Bryansk, Orel, Voronezh and Kaluga, in the so-called *Kursk Magnetic Anomaly*. These are rich ores with few impurities, but there are also immense reserves of lesser quality. Small deposits near Lipetsk and Tula are of historical importance only. Powdery haematites in the Kerch Peninsula, relatively low in iron but containing manganese and vanadium, require sintering before use, but they can be worked inexpensively by opencast methods, and go mostly to the Zhdanov steelworks and give plentiful phosphoric fertiliser as a by-product. A variety of ores occur in the Kola Peninsula and in Karelia, mostly poor in iron but easily mined and dressed, so the main constraint to their use has been poor transport. Transcaucasia depends on opencast working of cobalt-magnetite ores at Dashkesan in Azerbaydzhan for the Rustavi steelworks.

The Ural is one of the largest iron-ore-producing districts, with varied deposits lying in a broad belt along the eastern flank, mostly rich in iron and worked by both opencast and shaft methods. Several deposits contain natural alloys with titanium, vanadium and manganese, but some of the early important workings are now largely exhausted, like the 'Magnitaya Gora' at Magnitogorsk, so that new ones must be established, with a most favourable area around Orsk-Khalilovo.

Western Siberia has two main workings – in the Altay and in Mountain Shoriya – whose ores are rich in iron and generally easy to mine. Promising deposits as yet unworked await provision of adequate transport facilities. Particularly significant in Eastern Siberia are the two main deposits of the Angara-Ilim basin, because these ores are self-fluxing, can be worked by opencast methods and are easy to dress, and would be useful if a large iron and steel complex reputedly planned for the Angara basin were built. Most reserves in the Far East lie in the south of the region, already mined in the Malyy Khingan range, where the ores are not particularly rich but are easy to work and dress. The Aldan ore deposits, easier to develop on completion of the Baykal–Amur trunk railway, favour the establishment of heavy metallurgy in southern Yakutia.

There are some of the largest iron-ore reserves in the USSR in northern Kazakhstan, notably around Kustanay. These ores, mostly suitable for the basic steel process, can be easily worked, going principally to the Ural works. The Ata-Su and Karsakpay deposits in central Kazakhstan are well placed to serve the steelworks at Temir-Tau and Karaganda, though the latter deposits have yet to be opened.

Manganese comes overwhelmingly from Chiatura in Georgia and Nikopol in the Ukraine. For the future, Tokmak near Nikopol is important. Secondary sources occur in the Ural, at Dzhezdy in Kazakhstan, in Western Siberia and in the Far East. Chrome comes mostly from Aktyubinsk in western Kazakhstan and from around Sverdlovsk in the Ural. Nickel deposits are scattered, but the most important deposit is near Murmansk, with others near Chelyabinsk and Orenburg in the Ural, near Aktyubinsk in Kazakhstan and at Norilsk in Eastern Siberia. Cobalt occurs mostly in polymetallic ores in the Ural, Kola Peninsula and at Norilsk. Tungsten is scattered in many places, but the major producer is Tyrny Auz in the northern Caucasus, with the Iultin mine in the Chukchi Peninsula important and some significant producers in Central Asia. Tungsten–molybdenum ores occur in the Yakut ASSR and at Dzhida in the Buryat ASSR as well as in the Altay country and notably near Karaganda in Kazakhstan. Molybdenum is found by itself in Siberia, Kazakhstan and in Transcaucasia, though the ores often contain only a fractional percentage of the metal.

6.5 Non-ferrous metals

Of the non-ferrous metals, copper occurs notably in the Ural, but the most important workings are in the south near Orenburg at Gai and Blyava. Copper pyrites are worked in the Stavropol area of northern Caucasia and in the Altay, but far more important is copper in polymetallic ores in Kazakhstan (Kounradskiy, Boschakul) and in Uzbekistan (the major Almalyk producer), which together probably now supply over half the total output. There are also useful deposits in Armenia and at Dzhezkazgan in Kazakhstan, but a major deposit is being developed with Japanese aid at Udokan near Chita in Eastern Siberia. Copper–nickel ores make substantial contributions from Norilsk and the Kola Peninsula.

Lead and zinc, commonly found in polymetallic ores, come from several deposits in Kazakhstan and elsewhere in Central Asia. In Siberia, the Salair and Nerchinsk deposits are important as is that at Dalnegorsk in the Far East. Sadon in Transcaucasia is a significant producer and there are workings in Armenia and in the Ural. Transbaykalia and the southern part of the Far East region were formerly the sites of the main tin producers, but interest has now shifted to north-east Siberia and the Chukchi Peninsula at Iultin. Despite the exceptionally hostile environment, the importance of the metal makes the cost and exertion worth while. The main sources for aluminium are the bauxite deposits of the Ural, Kazakhstan and old workings near Tikhvin in north-west European Russia. Future supplies could come from deposits in Siberia near Kemerovo and Krasnoyarsk. Large potential resources are the nephelines of the Kola Peninsula, Siberia, Kazakhstan and elsewhere in Central Asia as well as Transcaucasia. Antimony is mined in Central Asia and won from secondary sources in the Ukraine. Magnesium comes from Satka near Chelyabinsk and Solikamsk near Perm, while magnesium salts occur in the Kara Bogaz Gol on the eastern Caspian shore and from lakes in Western Siberia. The main sources of mercuric ores are Khaidarkan in Kirgizia and Nikitovka in the Ukraine, but other deposits occur scattered in Siberia, northern Caucasia and the Ukrainian Carpathians. Gold comes from the Ural, Siberia and the Far East as well as Kazakhstan. The Aldan Plateau and the Lena and Kolyma valleys are major centres of placer workings. Silver is derived from the lead–zinc ores of Siberia, Kazakhstan and Central Asia. Since the 1950s a main centre of diamond workings, both in placer deposits and from pipes, has developed in the Yakut ASSR.

6.6 Minerals for the chemicals industry

Minerals for the chemicals industry are also common and scattered. The Khibiny area in the Kola Peninsula contains the world's largest apatite deposit that provides alumina as a by-product, but is usually considered as a source of fertiliser. Central European Russia has several useful deposits of phosphates, while others occur in Kazakhstan (Karatau), parts of the Ural and in Estonia and in north-western European Russia. An exceptionally large deposit of potassium salts lies at Solikamsk near Perm, but there are also important reserves in the western Ukraine near Kalush and Ivano-Frankovsk as well as in Byelorussia and Lake Inder in Kazakhstan, not to mention elsewhere in the Ural and in the Turkmen republic. Central Asia contains the main reserves of native sulphur, which has also been worked in the Volga basin near Kuybyshev. The Ukraine, including the Crimea, has partly explored deposits, which have additionally been recorded in northern Caucasia, the Far East and Kazakhstan. Sulphur pyrites have been widely worked in the Ural.

Natural soda deposits of vast proportions occur at Mikhailovka on the Altay fringe. Soviet resources of common salt are virtually limitless, with the large reserves lying on the Carpathian footslope in the Ukraine, in the Ural, Eastern Siberia and the Yakut ASSR, while there is also no shortage on the lower Volga. Mirabilite and other types of salt occur naturally in the great evaporating pan of the Kara Bogaz Gol on the Caspian shore, besides numerous salt lakes once part of the Aral Sea and in the lakes of the Kulunda steppe. Mica is found in several places in Siberia, with the main workings in the Aldan Plateau and at Mama, but the shores of Lake Baykal were formerly important. New deposits have been recorded in the Murmansk area and in Karelia. The Ural is still the main centre for asbestos production, but the Dzhetygara deposit near Kustanay is now also important.

Such an imposing array of natural resources must be a great comfort to Soviet economic planners and clearly give the Soviet Union a particularly significant world role as natural resources become increasingly in short supply.

6.7 Where to follow up this chapter

Information about Soviet mineral resources is diffusely spread through many statistical handbooks and trade journals. Among specific texts are Shabad, T., *Basic Industrial Resources of the USSR* (Columbia University Press, New York, 1969), and Shimkin, D. B., *The Soviet Mineral Fuels Industry, 1928–1958* (US Department of Commerce, Washington, D.C., 1962) and Shimkin, D. B., *Minerals – A Key to Soviet Power* (Harvard University Press, Cambridge, Mass., 1959), while Kowalewski, J., 'The Soviet Union's Struggle for Self-sufficiency in Metals', *Optima*, 9, pp. 209–15, remains valuable. A Soviet source is Gerasimov, I. P. (ed.), *Natural Resources of the Soviet Union: Their Use and Renewal* (Freeman, San Francisco, 1971).

Other discussions include Dienes, L. and Shabad, T., *The Soviet Energy System* (Wiley, New York, 1979), Melnikov, N. V. *et al.*, *Toplivno-energeticheskiye Resursy SSSR* (Moscow, 1971), and Probst, A. Yu. (ed.), *Razvitiye Toplivnoy Bazy Rayonov SSSR* (Moscow, 1968).

The most valuable statistical source, with annotation and explanation, is in the *Minerals Yearbook* (US Department of the Interior, Bureau of Mines, Washington, D.C., annual).

7

Soviet Industry – The Major Branches

Soviet policy to make the country one of the world's major industrial powers and ultimately to overtake the USA has had such a high priority that non-industrial sectors of the economy have sometimes seen it pursued at their expense. When the size of the country and its resource endowment (see Chapter 6) is considered, such an aim is perhaps not so unrealistic as it may seem at first sight. With the economy centrally planned and directed, the state has had the freedom to allocate its resources and priorities as it has seen fit. Under Stalin, development was pursued in an encapsulated, closed economy that sought the highest possible level of self-sufficiency. Since Stalin's death, the stringency of this approach has been relaxed, with a greater involvement in Comecon and even trade outside the *bloc* allowed a measure of encouragement.

The structure and location of industry in the Soviet Union reflect both the rapidity with which the economy has developed and the impress of regional planning concepts framed in a Marxist-Leninist philosophy and ruthlessly pursued. After the destruction of the Revolution and ensuing civil war, a new start had to be made, and consequently in little more than half a century the Soviet Union has gone through the gamut of industrialisation that took virtually twice as long in most of Western Europe. Between the wars, the concentration was on developing a basic heavy industrial structure in which consumer goods had an extremely low rating, but since the Second World War emphasis has shifted to the more sophisticated, technologically advanced sectors of industry. This has included extensive development of the chemicals industry and even some modest development of consumer goods. Nevertheless, compared with its peers, Soviet industry still shows a larger-than-usual emphasis on the heavier branches and a markedly weak representation of consumer goods.

A pervading influence on industrial location has been the planners' concern to implement the Marxist-Leninist tenet that the level of regional development should be equalised as far as possible across the country. On to this concept, during the 1930s, was also hung the strategically motivated idea of regional self-sufficiency. Such a planning policy also fitted well with a desire to reduce the burden on the inadequately developed transport infrastructure. In developing so-called 'complex regional economies' to a high level of local self-sufficiency, it has been necessary in many instances to locate industry in sub-optimum situations and to develop sub-optimum mineral and other resources. Nevertheless, the Soviet authorities have shown a ruthless ability to assemble workers and equipment at selected growth points against sometimes truly formidable physical difficulties.

During the 1960s a shift away from developing major projects in greenfield sites in the remoter regions appeared, with a renewed interest in the more accessible parts of the country that possessed an already existing infrastructure, and this may reflect an accelerating drift towards a more technologically sophisticated economy. Most Soviet geography textbooks give the impression that since the Revolution industrial development has been concentrated essentially in the eastern regions, where certainly some exciting projects have been, but the development of European Russia has also been impressive, if without the spectacular projects so favoured by Soviet propagandists. According to some calculations, almost three-quarters of fixed industrial capital and over 80 per cent of the value added by manufacture remain in European Russia: in this latter category between 30 and 40 per cent lies in the central industrial region around Moscow, the Dnepr–Donbass area of the Ukraine and in the Ural region. Well over half total national investment remains in European Russia west of the Volga, where about 60 per cent of the labour force is found.

Peter the Great encouraged foreigners to settle in Russia to develop new industries and crafts, notably metal-working, textiles, brick-making, and china and glass manufacture, though production was directed essentially towards government needs. With iron ore, charcoal and water power for furnaces and hammers, the Ural became the focus of a considerable iron industry, so that during Catherine's reign Russia grew into the largest iron producer in Europe. Powerful families helped by organising the labour of their serfs, like the 15,000 people working salt and bog iron on the

Stroganov estates at Solvychegodsk. Industry was most successful in the northern forest lands where serfs paid money dues rather than gave their labour-power on the land and were therefore able to devote themselves to manufacturing all kinds of articles for sale in the great fairs. There was marked specialisation: some villages concentrated on fine iron castings, others on iron tools or even knives and padlocks, while others made textiles and wooden articles.

One of the earliest modern factory industries was cotton textiles, using at first imported raw materials, but later drawing them from Russian Central Asia and where British capital and management were notably active. It had a vast domestic market for its cheap goods and grew notably around St Petersburg, in towns to the east of Moscow, at Narva in Estonia and even in Russian Poland. Falling raw cotton prices after 1830 and high protective tariffs had given it the initial impetus, so that by 1860 factory cottons surpassed peasant cloths and competed fiercely with linen.

Following the emancipation of 1861, there was an upsurge in industry, especially after 1870, when the inflow of foreign capital rose sharply. Besides Moscow, St Petersburg and the Baltic provinces (where German entrepreneurs and capital were active), a rapid growth came from Jewish entrepreneurs in Russian Poland, but there was also investment in the Ukraine, in the Volga towns and in the Caucasian oilfields, while the first interest was shown in industrial development in Siberia. By the mid-1880s heavy iron and steel-making, using coke smelting, was growing on the Donbass coalfield, drawing ore by railway from Krivoy Rog, while a similar but smaller industry was developing around Dabrowa in Russian Poland. Under such competition, deprived of its serf labour and short of charcoal, the old Ural smelting industry declined quickly. British, Belgian and French money was invested in heavy industry, while British and German investment was made in the beginning of an engineering industry, in which railway engineering was notably important. Foreign capital represented 90 per cent of the total in mining, 50 per cent in chemicals, over 40 per cent in metallurgy and just under 40 per cent in textiles.

By 1913 there were already five emergent manufacturing districts that accounted for over 80 per cent of national output and contained a similar proportion of the Russian industrial labour force, whose modest size can be judged from the fact that over 70 per cent of all employment was still in agriculture. The most marked industrial concentration was around Moscow, accounting for over a third of the industrially employed and the value of industrial output. This was also the main commercial focus and its importance lay in its nodal position near to all the great fairs that still dominated Russian commerce. Engineering was notably significant, with the large Kolomna railway works opened in 1862. Not far to the south lay the iron-making towns of Tula and Lipetsk. To the east in humid lowlands lay the cotton towns, with Ivanovo as the third major industrial employer in the country. With 65 per cent of all spindles and 71 per cent of all weaving frames, over 80 per cent of all textiles were produced here. Further east, Gorkiy (then Nizhniy Novgorod) built river boats, railway equipment and other engineering products, while well to the south-west lay the big railway works of Bryansk-Bezhitsa.

The Baltic towns and St Petersburg contributed about 15 per cent of industrial employment and output, mostly by working imported raw materials. Manufacturing was mostly for government contracts, but industries included electrical engineering, chemicals and rubber, heavy cloths, clothing and shoes. The German minority in the Baltic towns comprised an important entrepreneurial class, notably in Latvia at Riga, while at Narva in Estonia British management ran the biggest single industrial enterprise in the tsarist empire, the *Krengolm* mills employing 12,000 workers.

The Ukraine claimed a fifth of industrial employment and contributed well over a fifth of output. There had been rapid growth of heavy industry, often run by foreign overseers even if financed by Russians, with a major stimulus coming with completion of the Krivoy Rog–Donbass railway in 1884. By 1913 well over two-thirds of iron and steel output in Russia was from the Donbass and over half the production from heavy engineering came from the Ukraine, where there were the big Lugansk railway works and the shipyards at Nikolayev and Kiev. Over 80 per cent of Russian bituminous coal output at this time came from the Donbass. Although outclassed by the Donbass, the Ural still produced significant quantities of high-quality pig iron and there was firearms manufacture as well as foundry work in towns like Perm, Izhevsk and Zlatoust. Nevertheless, much of the rich and diverse mineral wealth of the Ural still remained untouched.

By 1913 the coming of the railway to Siberia had awakened an interest in development. Coal-mining had started in the Kuzbass, while coal was even being mined at Karaganda and Ekibastuz in Kazakhstan for a small smelter at Spasskiy Zavod. Gold-mining was largely under state control, but there were some foreign investments. Salt was also being worked in the Kulunda steppe, while in Baykalia small foundries were working at Guryevsk and Petrovsk-Zabaykalskiy. In the Far East, mostly for naval use, coal was mined at Suchan and some engineering done in Vladivostok, but foreign capital had opened lead and zinc mining at Dalnegorsk (Tetyukhe).

One of the most significant mining areas immediately before the First World War was Caucasia. Financed by the Nobel group, but also by others, the rich oil deposits were making Russia one of the leading world producers, while the valuable manganese of Chiatura was also being mined. Raw cotton was already beginning to flow from Central Asian territories not long incorporated into the tsarist empire, and around the eastern Caspian the resources of oil and salts were being looked at critically.

The Revolution and subsequent civil war spread destruction through existing industry and important plants in Poland, Finland and the Baltic republics were lost by secession. By the early 1920s a new start was necessary, but there was no foreign capital and little outside help. Conditions were uncertain and included the brief *New Economic Policy* (NEP), a partial return to private enterprise, but new ideas were emerging and state control of all the means of production was reintroduced as soon as practicable. By 1927 production levels were back to 1913, though they were low even then by Western European standards. If the Soviet state were to survive and to strengthen its political and economic position, massive industrialisation was imperative, but because of the hostility of the world outside it was felt vital to develop this industrial system in a closed and overwhelmingly self-sufficient economy. The new directions were well seen from the start of the first Five-Year Plan in 1928. Expediency demanded that industry in already established areas should be rehabilitated first and existing pools of skilled labour be kept *in situ* to train further cadres. Essential equipment not obtainable at home was bought from abroad from resources sorely needed at home; farming had, for example, to export much-needed foodstuffs to pay for industrial machinery imported under the greatest priority. Some grandiose plans were formulated, though few were carried through, but the broad general strategy was extensive electrification under the Goelro plan, while a pre-1913 idea of linking Ural iron ore and Kuzbass coal in a massive interchange was revitalised by development of the *Ural–Kuzbass Kombinat*, to be a second metallurgical base to bolster the first, the Donbass. The first stages were realised between 1924 and 1932, with the Magnitogorsk steelworks opened in 1931, but lying several hundred kilometres apart and dependent on inadequate rail links the scheme was never a great success. With discovery of iron ore near the Kuzbass and the opening up of coal resources closer to the Ural in the Karaganda basin, the emphasis shifted to independent development of the Ural and Kuzbass.

Marxist-Leninist doctrine stressing equalisation of regional development stimulated interest in the eastern regions, reflected in the *Ural–Kuzbass Kombinat*, the opening up of the Karaganda coalfield and the strategically motivated Komsomolsk steelworks in the Far East. Siberia was also secretively remote, appealing to strategic interests and to the planners' concept of a closed economy, and its resources fitted the emphasis on heavy, basic industries producing capital goods with which to build more heavy basic industries. The Second World War gave a further impetus to the strategic shift of industry into the eastern regions as plants were hurriedly moved to escape the invading Germans. The wisdom of investment in plants east of the Volga was now justified as towns in the Volga, Ural and Siberia became the industrial mainstays of the war effort. Soviet sources claim that the Second World War retarded economic development by at least eight years, but it is clear that the eastern regions emerged in 1945 vastly more developed than they would otherwise have been, had not the frantic shift of all sorts of works taken place in 1941–2. Few of the evacuated plants were to return to their old sites in European Russia, which were usually completely re-equipped.

By the end of the 1940s industrial output was back again at the 1939 level. Under Stalin, little change was made to economic strategy in the post-war years and in a way rehabilitation of wartime damage called for the same approach as after the civil war, with emphasis on heavy capital goods for reconstruction. However, the process was this time boosted by valuable reparations from Germany and the ruthless exploitation of the Eastern European satellite economies. After Stalin's death in 1953 there was a slow change, and as life was breathed into Comecon the Soviet authorities became less obsessed with concern for a closed and self-sufficient economy. This change was certainly caused by a quiet but relentless internal pressure for better living standards, forcing the release of resources from capital to consumer goods, while there was influence from the Eastern European members of Comecon to open a more equitable pattern of trade and economic relations between them and the USSR. At the same time, there has been increasing discussion of the best ways of organising the economy, with ideas swinging between decentralisation (as under Khrushchev) to a return to a more centralised form (under Brezhnev), though neither seems to have reached the desired goal.

Industrialisation itself is changing, with consequent impact on spatial patterns of its distribution. There is in all branches a growing sophistication and a shift from relatively bulky, simple basic capital equipment to more highly technological installations and machinery requiring higher levels of skill. Labour itself is becoming less easily available as the demographic structure changes, and as demand for and expectancy of higher living standards grows it is less easy to attract people into the harsher corners of the country. On the other hand, some of the more easily accessible resources are being exhausted and development has to turn to the more inaccessible parts of the country, with consequently higher costs and greater challenges. In themselves, these changes have made reconsideration of development trends in and between the major planning regions necessary, reflected in renewed interest in industrial investment in some long-neglected parts of European Russia, like Byelorussia and the western Ukraine.

7.1 The metallurgical industries

After the Revolution, industrialisation gave a particular stimulus to iron and steelmaking as a supplier of a basic need for economic growth. Expansion of the industry was favoured by the plentiful reserves of coking coal and iron ore as well as alloy metals. Although it was felt desirable to follow the planners' principle of spreading industry across the regions, it was also felt necessary to speed development by concentrating the industry on a few favoured districts, where valuable economies of scale could be achieved, and the concept of the 'metallurgical base' was promulgated. The renovated works of the Donbass, with its sound resource endowment, formed the first of these bases. By the mid-1920s the Ural was under development as the second, but because good coking coal was poorly available it was to be associated with the Kuzbass coalfield, where at that time no useful iron-ore resources were known. This development was framed in a concept of a gigantic industrial *Kombinat*, whose constituent units, the Ural and the Kuzbass, were separated by several hundred kilometres linked by the Trans-Siberian railway and satisfying the gigantomania then pervading Soviet thought.

Planning was interrupted by the war, but by the early 1950s a third metallurgical base was being discussed. This arose from the weakening in the mid-1930s of the Ural–Kuzbass link as iron ore had been discovered in Western Siberia and the coking-coal deposits of Karaganda had been developed for the Ural. The new third metallurgical base, spread from the large works being built in the Kuzbass across southern Siberia into Transbaykalia, was to draw coking coal from the Kuzbass and from southern Yakutia, while iron ore was to come from deposits including those of Mountain Shoriya and the remote Angara-Pit basin, as well as from the Aldan Plateau. The original plan included a large iron and steel complex at Tayshet and possibly also further east on the Trans-Siberian railway. In recent years this plan appears to have been shelved.

There were, however, many scattered plants built as part of the over-all planning aim to spread industry among the major regions. Some of these, like the Komsomolsk works in the Far East, were generated by strategic considerations of a resurgent Japan in the late 1920s. This is a full-cycle plant of about 1 million tonnes annual capacity and using both iron ore and coking coal available regionally. In Central Asia during the late 1930s an above-average supply of scrap was generated and to use it the Bekabad (Begovat) steelworks was built, becoming operational in the 1940s. It also augmented the production of the big full-cycle producers using good local coking coal and iron ore at Karaganda and Temir-Tau in Kazakhstan. Other small regional producers include the Liepaya plant in Lithuania, the Red October plant at Volgograd, and the renovated steelworks and rolling mill at Petrovsk-Zabaykalskiy east of Lake Baykal. In Transcaucasia, the Zestafoni alloy steelworks uses Chiatura ores, while Rustavi produces a range of products, and the Sumgait plant (Baku) serves the Azerbaydzhan oil industry.

With its large engineering industry, it is not surprising that several important steel plants are found in the central industrial region, mostly concerned with scrap conversion and short runs of specific steels, like those in Moscow and the Elektrostal plant at Noginsk. Unlike some regional producers that have to augment scrap by bringing in long-haul cold pig, there is in this cluster also an iron-making capacity, notably older plants at Tula, Orel and Lipetsk besides the Kosaya Gora ferro-alloy works. Since 1952 a large full-cycle plant at Cherepovets on the Rybinsk reservoir has been using Karelian ore and Pechora coal to supply the central industrial region and the north-west, which has small works at Kolpino and Vyartsila, as well as Leningrad.

The development of the ore resources of the *Kursk Magnetic Anomaly* will enhance metallurgy in central European Russia. A large full-cycle plant is to be built with foreign assistance at Staryy Oskol and there will be expansion of capacity at Orel and Lipetsk. This new 'territorial production complex', along with the expansion of the capacity of existing and additional plant in the Kuzbass and in Kazakhstan, seems to have contributed to postponement of the third metallurgical base.

Over 70 per cent of all iron and steel semis are produced in the Ukraine and the Ural. Ukrainian heavy industry lies on the Donbass coalfield and in the bend of the Dnepr river, so that both areas can obtain good coking coal from the Donbass mines and excellent ore from Krivoy Rog. There is also a supply of ore suitable for Thomas steel from Kerch in the Crimea, and the vast resources of the Kursk deposits will also be easily used in the Ukraine. Both limestone for flux (from Yelenovka and Karakub) and manganese ore (from Nikopol) are available. It is a primary railway focus, making distribution of its products easy to all parts.

The main steel towns on the coalfield are Makayevka, with its massive integrated plant and pipeworks, and nearby Donetsk, which also has a large plant. Together these two centres smelt about half the metal produced in the Donbass. But there are also Kramatorsk (electric steels and steel conversion), Konstantinovka (smelting and rolling), Kadiyevka (another full-cycle plant), Kommunarsk, Voroshilovgrad (rolling mills) and several smaller works. All these are plants at the western end of the coalfield, nearest to their long-standing ore supply from Krivoy Rog. In the east, in the Rostov *oblast* of the RSFSR, there are works at Krasnyy Sulin and Artemovsk, both with big rolling mills. Away from the coalfield, on the Azov coast, Zhdanov has a big full-cycle plant, using Gorlovka coke and mostly Kerch ore. It sends metal to rolling mills at Kerch.

Krivoy Rog is still an important ore producer, despite now having to work lower grades, and fragile ores are smelted locally, with Donbass coal and coke. Both Dneprodzerzhinsk and Dnepropetrovsk smelt Krivoy Rog ores with Donbass coke, though the latter town specialises in open-hearth conversion. There are rolling mills at Novomoskovsk and Nikopol, while Zaporozhye smelts and rolls. With considerable local reserves of manganese lying between the Dnepr and the coalfield and with other alloy ores imported from the Ural and Transcaucasia, there is a substantial ferro-alloy industry. Kadiyevka, using thermally generated electricity, is the main plant on the coalfield, while, using hydro-electric current, major plants operate in Zaporozhye and Nikopol.

The Ural came back into prominence with development of the second metallurgical base in the inter-war period. Underlying iron and steelmaking in this region is the generous endowment with high-quality iron ore and many alloy metals, as well as a good supply of fluxing materials. The main iron and steel towns are situated mostly along the highly metalliferous belt along the ranges' eastern flanks. The major drawback arises from the lack of adequate deposits of good coking coal, which has to be brought in either from the Kuzbass or from Karaganda and related deposits. Kazakhstan is also beginning to send ore to the Ural as some of the local reserves are exhausted. The main export from the region is rolled steel in various qualities and forms, with over a third going to the central industrial region and about a quarter to Siberia.

There are six main groups of plants. Serov, the most northerly, uses Kuzbass coal and coke and local ores, though ore is now also brought from Kachkanar, 150km away to the south-west. A ferro-alloy plant uses manganese from Polunochnoye and there is considerable specialisation on special steel, notably at Severouralsk. The Nizhniy Tagil group has one of the largest fully integrated iron and steelworks in the USSR. Again, coal and coke come mostly from the Kuzbass, but depletion of local ore resources has meant importing from other mines, particularly from Kachkanar.

Sverdlovsk has a cluster of important works, though it has no fully integrated plant. Concentration is on special steels, notably for electrical engineering, using pig iron from Nizhniy Tagil and from Magnitogorsk. To the south lies Chelyabinsk, where steel was first made in 1943 and blast furnaces built after 1945, so that the town now has one of the larger fully integrated works. Ore comes from Bakal and coal and coke from both Kazakhstan (Karaganda and Ekibastuz) and the Kuzbass, while ore has begun to be imported from Rudnyy in Kazakhstan. The local lignites fire large thermal power stations whose current is used in electric steelmaking (notably ferro-alloys) at Chelyabinsk and Zlatoust.

Magnitogorsk, an industrial complex developed in the earlier five-year plans, originally used ores from Magnitnaya Gora just on the edge of the town, but it now depends increasingly on ore and coke from Kazakhstan and flux from the Agapovka limestones. The works that dominates the town is the largest single producer in the Soviet Union with probably about 15 million tonnes a year. On the south, the Novo-Troitsk works, a medium-sized, fully integrated installation, opened in 1945 near Orsk to use local ore and also ore from Rudnyy in Kazakhstan as well as Karaganda coal and coke.

The western slope of the Ural has a few small plants, mostly in the upper Kama valley. The most important is an integrated works at Chusovoy, using local Kizel coal for power but coke from other sources as well as local ores, and another significant plant lies at Lysva, both associated with the Nizhniy Tagil complex.

The history of the iron and steel industry in the Soviet economy has savoured of the situation during the nineteenth-century industrialisation of Western Europe. Now, however, with the growing sophistication in the product range of Soviet industry and the move towards completion of the basic economic infrastructure, heavier industries like iron and steel will doubtless begin to play a less significant role in the over-all economic scene, in the way that has occurred in the West. This appears to be reflected in the quiet shelving of the Siberian third metallurgical base through a shift in interest to the *Kursk Magnetic Anomaly* ores and the building of one new works at Staryy Oskol, trends that suggest the industry has reached a plateau. Planned increase in the output of steel in various forms has been modest compared with manufactured goods, so that the additional production can be achieved using modern technology to boost the productivity of existing plants. In fact, as in the West, rising output may be achieved using fewer plants. There is plentiful evidence to show that Soviet interest is increasingly concerned with raising plant and worker productivity. Furthermore, existing plants near declining high-grade ore sources, which a few years ago had a bleak future, have been given a new lease of life by methods that allow use of much lower grades of ore.

If the growing involvement of the Soviet Union in world trade and the broad spectrum of world industry continues, another problem may arise for its iron and steel industry. The most successful plants in the West are those with an annual capacity of ten or more million tonnes per annum, highly automated and using the most modern technology, making use of massive inputs of raw materials transported at the lowest possible unit cost by bulk ocean-going carriers. The effect of this has been to force *inland* producers to seek comparable locations, notably on high-capacity waterways with easy access to coastal ore and coal terminals. The coastal type of location is conspicuously rare in the Soviet setting and even appropriate inland riverine sites would be difficult to find. Overwhelmingly dependent on rail transport, the transport problem presents one of the most serious constraints to Soviet hopes of taking advantage of the economies of scale possible in the 'Japanese-style' coastal steelworks.

The most promising plants for long-term development appear to include at the present top end Cherepovets (to be raised to produce 10–15 million tonnes per annum), Krivoy Rog (to produce over 10 million tonnes), Magnitogorsk (already planned to produce 20 million tonnes per annum by 1985), Novo-Lipetsk (*circa* 10 million tonnes), and the plant at Staryy Oskol. Possible other sites for expansion are Nizhiy Tagil (presently over 6 million tonnes), Chelyabinsk (also over 6 million tonnes), Zhdanov (over 5 million tonnes), and the plants in the Kuzbass, the Kazakh works at Karaganda and Temir-Tau, besides the Ukrainian works at Dneprodzerzhinsk, Zaporozhye and works currently being modernised at Donetsk. All have particularly favourable transport facilities and sources of raw materials or are especially well placed for major customers.

7.2 Non-ferrous metallurgy

The location of non-ferrous metal smelting is both close in many instances to deposits and in some significant cases well away from any of the mines. The siting of smelters and refineries at or near mines depends on the nature of the deposit in terms of size and expected life as well as its remoteness. There is also the transport factor: carriage of refined metal is generally more worth while than movement of heavy, bulky ore, of which a considerable proportion is ultimately waste. Copper refining and smelting expanded greatly between the wars and the main plants have been in the Ural, near to raw ore, at Krasnouralsk, Sredneuralsk, Karabash, Revda and Mednogorsk, but much of their ore now comes from Gay, Uchaly and Sibay in Bashkiria. During the 1930s the Balkhash smelter using Kounradskiy ores was built, and later with opening of the major Dzhezkazgan deposits the Karsakpay smelter was greatly expanded. The Kafan, Alaverdi and Kadzharan mines have their ores refined and smelted locally in Armenia. During the 1960s the older Ural mines' output has declined, whereas those of Kazakhstan have increased, notably new mines like Bozshakul, Chetyrkul and Sayak, so that the dominance of the Ural smelters is also threatened. Copper smelting of by-product ore also takes place at Monchegorsk and Pechenga in the Kola Peninsula, at Norilsk and at mines on the Altay flanks. In 1964 a major smelter was opened at Almalyk in Uzbekistan and new smelters are planned for Urup in Caucasia and Udokan in Siberia. The most unusual location is Moscow, where a smelter prepares special pure metal from ingots brought from other smelters.

Lead and zinc commonly found together in ores are separated generally at or near the mine, with the zinc usually left as a concentrate to be sent to refineries near to large supplies of electricity needed to produce metallic zinc. The original plant of pre-revolutionary times at Ordzhonikidze (Vladikavkaz) still works, turning out metallic lead and zinc, and also silver. The main lead smelters are now situated at Chimkent, Leninogorsk and the Far East Dalnegorsk (Tetyukhe) plant. A large zinc plant lies at Belovo on the Kuzbass coalfield, while other plants include Ust-Kamenogorsk in Kazakhstan, Chelyabinsk in the Ural and Konstantinovka in the Donbass. A large modern electrolytic zinc plant has been opened at Almalyk in Uzbekistan.

A key branch of non-ferrous metallurgy is aluminium manufacture, where the continuing expansion of productive capacity reflects its importance to the Soviet economy. The original modest endowment with conventional ores like bauxite has been widened by ways of using alternatives like alunite and nephelite. Considerable quantities of bauxite are, however, imported from Greece, and Hungary sends alumina. The industry is characterised by reducing raw ore to alumina, which is then converted into metallic aluminium, a stage that requires immense quantities of electric current, whose availability becomes a locational key. The clear staging of production processes results in some long-range regional links between plants.

In the inter-war period bauxite from Boksitogorsk, south-east of Leningrad, was turned into aluminium at an early hydro-electric plant on the Volkhov. At this time large deposits of bauxite discovered in the northern Ural were worked at Severouralsk, while in 1939 a processing works opened at Kamensk-Uralskiy. Ore from Boksitogorsk was also processed at Zaporozhye near the big Dneproges hydro-electric station. The war years saw plants evacuated to the eastern regions, with a new plant built at Krasnoturinsk in the Ural and another at Novokuznetsk in the Kuzbass. After the war, the Boksitogorsk–Tikhvin and Volkhov industry has been expanded, augmented by a new works at Pikalevo, but they now depend increasingly on by-product ore from the Kola apatite, though a large bauxite deposit in the Onega basin is to be developed. Some alumina from these plants goes north to Nadvoitsy and Kandalaksha, where there are electrolytic refineries.

Greek bauxite, of the order of half a million tonnes a year, is used at the Zaporozhye plant and Hungarian alumina at Volgograd. Plans for other works on the Volga have been made. Local raw materials and hydro-electricity supply the Transcaucasian plants at Sumgait and Yerevan, which also take alumina from the Ural works. In the long term Siberia is likely to become a major aluminium producer as its vast hydro-electric potential is developed. In the 1960s important plants were opened at Shelekov near Irkutsk and the Angara barrage, at the hydro-electric station at Bratsk and at Krasnoyarsk near another hydro-electric barrage. Along with the older plant at Novokuznetsk, these now produce probably over 50 per cent of Soviet aluminium. Unfortunately, large quantities of alumina for these plants have to be brought from the Ural and from Kazakhstan (where it is produced in a plant at Pavlodar), but Siberian alumina production will be increased by the new Achinsk plant using Kiya-Shaltyr or Belogorsk nepheline. Plans are also in hand to use kaolin sources in Central Asia, for which a plant is being built near Dushanbe.

7.3 Engineering

The engineering industry produces a wide range of diverse articles from a great selection of raw materials, mostly but not exclusively metallic. Although labour as a component of cost varies from branch to branch, it is nevertheless a major element, so its availability in quality and quantity is a key locational factor. To this must be added accumulated skill and design and development facilities. Because most engineering products stand transport costs over appreciable distances, thus considerable savings in unit costs can be made through mass production and by locational association of related branches. Even so, engineering is market-oriented and producers tend to lie relatively close to their principal customers (e.g. textile-machinery manufacture in textile-manufacturing districts), while some sections have special site considerations (e.g. shipbuilding).

Engineering in some form or other is found in almost every Soviet town, though much of the industry comprises small and relatively inefficient plants. Reputedly over half the value of production is contributed by less than 10 per cent of the plants. Some so-called 'engineering' plants are in effect simply repair and maintenance shops, while elaborate linkages between plants exist where components from different producers are assembled to produce the final product (e.g. the motor-car industry). Soviet sources give, however, little information by which the relative importance of either individual plants or individual towns may be judged, but from the scant information available the principal centres as well as plants can be reasonably identified.

Moscow stands out as particularly important in the national scene, both through the number of plants and their diversity, while other towns of European Russia like Gorkiy, Yaroslavl, Kalinin, Kolomna and many more make the central industrial region and the immediately adjacent Volga valley a notable focus of engineering. Another commanding cluster lies in the north-west around Leningrad, while some significant plants are situated in the Baltic republics, most notably in Riga. Engineering is also important in the Ukraine, with the Donbass and Dnepr bend towns associated with heavy engineering and centres like Kharkov, Kiev and Lvov with more specialised branches. The Ural has heavy engineering plants, much augmented during the war years by plants brought from further west in the country, while since the late 1950s the Volga valley has seen a formidable growth, with the opening of the big Tolyatti motor works and the large lorry plant on the lower Kama at Naberezhnyye Chelny. Smaller and more scattered clusters of works can be found across Siberia, notably around Novosibirsk and on the Kuzbass coalfield, while in both Central Asia and Kazakhstan as well as Transcaucasia there are works of national renown.

Building and mining equipment and machinery for the iron and steel industry is chiefly found in the Ural and Donbass and to a lesser extent on the Kuzbass coalfield, since equipment of this type is usually designed for a specific task for a particular customer. The Donbass plants are notably at Gorlovka, Novokramatorsk and Voroshilovgrad, while, some distance away, Kharkov makes mining equipment. In the Ural, Votkinsk, Sverdlovsk and Kopeysk are representative centres, and on the Kuzbass coalfield there are works in Prokopyevsk and at Anzhero-Sudzhensk in the north. In Eastern Siberia works provide specialised machinery for gold-dredging and diamond-working. It is not surprising that apparatus for oil-working comes especially from Transcaucasia (Baku) and from the towns of Armavir and Groznyy, while Ufa-Chernikovsk supplies the Ural–Volga oilfields.

Machine-tool manufacture is widely scattered, but the central industrial region's towns supply at least a quarter of total output. Other major plants lie in the Central Black Earth region and in the Ukraine, notably Kiev and Kharkov, while Leningrad is also a main supplier, all districts where there are ample design and development facilities, and Minsk in Byelorussia and the Siberian capital, Novosibirsk, are significant nationally. Textile machinery comes chiefly from the textile towns east of Moscow, whereas electrical equipment is designed and made notably in Moscow and Leningrad, while assembly takes place in many scattered centres. Newer plants lie in Siberia (e.g. transistor radios assembled in Novokuznetsk), Central Asia, Kazakhstan and Caucasia. Sverdlovsk and Novosibirsk are known for heavy electrical equipment. Much of the mainstream radio, television and electronics industry lies concentrated in the towns of the central industrial region and along the Volga. Kharkov was chosen in the 1930s to make the first Soviet cameras modelled on German *Leicas*, but during the war optical and related precision engineering spread to include towns like Kurgan, Tomsk and Zlatoust, while Cheboksary now makes film equipment. Power-generating installations – boilers, turbines, etc. – are made in Moscow and Leningrad and notably in Taganrog, but there are several other scattered works. Moscow, Saratov, Kuybyshev, Sverdlovsk and Minsk seem to dominate the manufacture of ball-bearings and related goods. Agricultural machinery is made throughout the country, but the manufacture of combine harvesters takes place generally near to the main areas where they are used, as at the huge *Rostselmash* works in Rostov on the Don, while tea-picking equipment is produced near the tea gardens of Transcaucasia, and in Central Asia cotton pickers are made.

There is a particularly important transport engineering sector, the basis of which dates back before the Revolution. The railways in tsarist times built a lot of their own equipment, so at several focal points in the network works were established, often with German help. After the Revolution railway locomotives and stock were built at large plants at strategic points in the network, but the switch to diesel and electric traction closed some of these plants.

Kharkov, Voroshilovgrad and Kolomna are now the main producers of diesel locomotives, but electric locomotives are notably made at Novocherkassk and Tbilisi. Ulan Ude in Eastern Siberia is a major supplier to Siberian railways. Coach and wagon works are sited all over the system, though some of the biggest are in the railway directorates that load or unload most wagons, so a region like the Ural shows a marked concentration. Railcars and metro vehicles are made principally at Mytishchi near Moscow and in Riga.

Motor-vehicle building was developed between the wars with American help and capacity further increased after 1945 by equipment dismantled in Germany. During the 1960s, again with foreign help, a big expansion took place. The early works were at Moscow and Gorkiy, while the Yaroslavl works had been brought from Riga in 1917. During the war years, lorry and heavy-vehicle building was expanded, notably in the Ural at Miass and on the Volga at Ulyanovsk. The works in post-war years were augmented by light vehicles (minibuses) built in Riga, dump trucks and heavy lorries from Minsk and buses from Zhodino (Byelorussia) and Lvov (Ukraine), while light motor-cars come from Zaporozhye and other vehicles from centres such as Odessa, Kremenchug and Melitopol. Kutaisi in Georgia builds lorries and there is also assembly at Petropavlovsk in Kazakhstan. Two very large plants developed in the 1960s are the *Fiat*-type plant at Tolyatti on the Volga making the successful *Zhiguli-Lada* range of vehicles and the huge Naberezhnyye Chelny *Kamaz* lorry plant. A common complaint has been the lack of major vehicle assembly plants in the eastern regions, though Kurgan, Omsk and Novosibirsk do this type of work. At Rubtsovsk is the Kharkov tractor plant evacuated during the war. Other big tractor plants are at Volgograd and Chelyabinsk, both wartime builders of armoured fighting vehicles. The clear grouping of motor-vehicles in Central European Russia and the Volga as well as the southern industrial area suggests the importance of what are essentially assembly plants being reasonably close to where most component manufacturers are and within the main market sphere as reflected by population distribution, road density and other transport statistics.

Shipyards for river vessels are found on each of the main rivers, though the largest and busiest are on the Volga, notably at Gorkiy and Krasnoarmeysk, and, at the delta, Astrakhan builds ships for both the Volga and the Caspian, which is also supplied from Kaspiysk. Sea-going vessels are built or maintained in yards on each of the Soviet seas. On the Baltic the main yards are at Leningrad and smaller yards in the Baltic republics. For the northern sea route both Murmansk and Arkhangelsk have yards. The Black Sea yards are among the biggest in the Soviet Union, notably those at Nikolayev, but others lie at Odessa, Kherson and Osipenko. On the Pacific coast Vladivostok, Petropavlovsk-Kamchatskiy and Nikolayevsk have repair yards, but no actual building is undertaken in the Far East.

Soviet sources give little information about aircraft construction, even though the country is now the largest producer after the USA. All signs point to pools of skilled labour and component assembly as well as design and research facilities lying in European Russia, notably in the central industrial region around Moscow, but also perhaps in the north-west around Leningrad and in the Ukraine. There may also be a considerable strategic presence in Siberia, while the 'cosmodrome' at Baykonur in Kazakhstan no doubt has associated plants not far away. Its location seems to have been chosen for its secrecy deep in the interior and also for its gravitational advantages so far south and for the open steppe as a suitable retrieval area.

7.4 The chemicals industry

A small chemicals industry, based mostly on imported raw materials, existed before the Revolution. A considerable development was made in early Soviet times, but by the later inter-war period the industry had fallen well behind the outside world. Little was done to improve its performance immediately after the war, apart from strategic needs, but a new impetus came from Khrushchev in the mid-1950s, concerned with the possibilities of petrochemicals but even more so with the need to develop an adequate production of artificial fertilisers and pesticides to support the new agricultural campaign.

The chemicals industry uses a wide range of raw materials, both mineral and vegetable as well as items arising as waste from other industries. Consequently location for many plants is orientated towards raw-material supply, especially where there is considerable weight loss in the process of conversion. In some instances the finished product is hazardous to transport, so manufacture is best undertaken near the consumers, commonly other industries, consequently giving rise to elaborate linkages.

Because of its diversity, Soviet planners have found it even more difficult than in other industries to provide complete regional self-sufficiency in chemicals products, so that substantial interchange between the regions takes place. The central industrial region around Moscow has a peculiarly significant position as the main focus of the more specialised branches, like pharmaceuticals, pesticides, cosmetics and films. There was an early presence here of the manufacture of synthetic resins, plastics and yarns. A smaller but important cluster lies in and around Leningrad and includes some of the oldest works, like the Red Triangle rubber plant. Chemicals works are widely scattered through the Ukraine, but a notable cluster, with the emphasis on essential alkali and acid products, lies on the Donbass coalfield, where coal provides the raw material for benzol, tars, and so on. A similar, though locationally more diffuse, pattern is found in the Ural, which benefited from plants evacuated there during the war. The oil and gas industry in the Ural–Volga oilfield as well as in Caucasian fields has seen the growth in these districts of petrochemicals associated with local refineries. Some petrochemicals plants are also found on other fields, like detergent-making in Krasnovodsk and carbon-black production at Ukhta. The chemicals industry of the lower Volga makes considerable use of salts from deposits like Lake Baskunchak or Elton or from the Kara Bogaz Gol. Caucasia uses its oil and gas resources but also its limestone and by-products from non-ferrous metal production. The mining activities of Central Asia and Kazakhstan also provide a similar base for the chemicals industry, with an important contribution made by artificial fertilisers. The development of petrochemicals using piped oil and gas has also occurred in southern Siberia in towns like Novosibirsk, Kemerovo and Krasnoyarsk as well as Angarsk and Barnaul, with manufacture of various plastics and synthetic yarns. The long-term energy situation in Siberia favours expansion of electro-chemicals.

A typical locational feature marks such an essential substance as sulphuric acid. Until recently iron pyrites has dominated the raw-material base, notably from the Ural, but with growing sulphur mining in Central Asia (e.g. at Gaurdak) and near Kuybyshev on the Volga as well as on the Carpathian footslope, interest has shifted to this raw material. Nevertheless, both pyrites and sulphur are sent to manufacturers near to the principal consumers, since the acid is hazardous to transport. As 40 per cent of the acid is used in phosphate fertiliser factories, production is close to such plants, which are notably grouped around Moscow and Leningrad and in the Ukraine. The phosphate rock has generally come from the Volga basin and the Ukraine, but the principal supply is now from the Kola Peninsula. Non-ferrous metal working gives possibly a quarter of the sulphuric acid produced, with the main plants at Alaverdi in Transcaucasia and Almalyk in Central Asia, but also at Revda and Krasnouralsk in the Ural. A thermal process is used to convert phosphate rock from the Kara-Tau deposits in Central Asia to fertiliser in plants at Dzhambul and Chimkent that use natural gas. Nitrogen fertiliser production is more complex and the early process derived hydrogen from coal, so that plants gravitated towards the coalfields, notably the Donbass, the Moscow lignite field, the Kuzbass and at Rustavi in Transcaucasia, while a similar base was used at Berezniki in the Ural. Nowadays natural gas is used in the ammonia process and plants have been converted to use this source on their old sites, but there are also new ones at Grodno in Byelorussia, at Rovno in the Ukraine and at Kokhtla-Yarve in Estonia. Potash fertiliser production is closely related to potash deposits, principally from the Ural at Solikamsk and Berezniki, with the balance largely from deposits in the western Ukraine near Lvov. About half this fertiliser will come by the mid-1980s from the vast Soligorsk deposit in Byelorussia.

For alkali production availability of salt, limestone and cheap energy is critical. In this way Donetsk, Slavyansk and Lisichansk, using Artemovsk salt and local limestone plus available energy in the Donbass, account for about a third of soda output, while most of the remainder comes from the Ural at Berezniki and Sterlitamak, using raw materials from Solikamsk and Sol-Iletsk. Caustic soda also comes from the Donbass and Ural, but with a swing to electrolysis of salt (which also yields chlorine, another essential chemical) production has shifted to where large supplies of current are available. Consequently caustic-chlorine plants have been developed in the Volga basin at Kuybyshev and in Siberia near Irkutsk, drawing in this latter case salt from Usolye-Sibirskoye. Using cheap electricity based on Ekibastuz coal and Tavolzhan salt, a plant has been built at Pavlodar in Kazakhstan, and also one at Yavan in the Tadzhik republic to use Nurek hydro-electricity, while other plants lie in the Crimea and the Altay country.

In the search for self-sufficiency the Soviet Union became one of the earliest manufacturers of synthetic rubber in the 1930s. The first plants using potato alcohol were near Leningrad, Yaroslavl, Voronezh, Tula and Kazan. Using local limestone for acetylene, a plant was opened in 1940 at Yerevan. After the war production shifted to synthetic alcohol and by-products of petrol refining, with the first plant at Sumgait near Baku, while as improved techniques have been used manufacture has shifted closer to oil refineries, with new works in the Ural–Volga oilfields and old works everywhere converted to the new processes. In Siberia the readily available wood alcohol is still used at Krasnoyarsk and an older acetylene plant runs at Usolye-Sibirskoye. In general, the massive expansion of oil and gas production during the 1960s has been reflected in increasing output of chemicals from a petroleum base. Early developments grew from thermo-plastics, mostly phenol-based from coal, though urea resin (from ammonia and made at the nitrogen fertiliser plants) went into production in the 1950s. Subsequently polyvinylchloride output was expanded as more chlorine became available, notably at Novomoskovsk, Sterlitamak, Usolye-Sibirskoye and Yerevan. Polyolefin resins have been increasingly manufactured near to refining centres.

Artificial fibres provide a bridge between the chemicals industry and textile manufacture. Acetate and viscose fibres were first made, with wood cellulose used, in small pre-war viscose plants at Klin, Leningrad and Mogilev, while acetate fibre was produced in Kalinin and near Ivanovo, all in traditional linen or cotton textile areas. Production in the Ural, Central Asia and Kazakhstan represents plants evacuated there during the war years, but since then additional plants have been built in European Russia, besides Barnaul, a Siberian textile centre, and at Krasnoyarsk. The real expansion has come in the synthetic fibres of the nylon–terylene type, with plants near coal-based chemicals (Moscow lignite field, Kuzbass) or where there are oil refineries or piped supplies of oil and gas.

7.5 Textile industries

Textile manufacture remains overwhelmingly in areas where it was established before the Revolution as a factory industry. Over three-quarters of all cotton textiles are still made in the thirty-odd towns in the humid lowlands east of Moscow, and the large textile mills at Narva in Estonia still operate. Even though large mills have been built in the Soviet period in Central Asia and in Transcaucasia, their contribution to the total appears small. Some cotton textiles are made in Siberia, but there has been a spread to towns in the Ukraine and in the Baltic republics. Woollen textiles also remain in the main pre-revolutionary centres, like Chernigov, Sumy, Klintsy and Bryansk as well as in and around Moscow, though new centres like Sverdlovsk in the Ural and Dzhambul in Central Asia have been added and the old works in Omsk are still operational. Linen mills remain concentrated in north-west European Russia, but there has been considerable encouragement in the Soviet period to the development of a factory-based silk industry in Central Asia (Osh, Samarkand, Tashkent, Dushanbe) and in Transcaucasia (notably Gori and Nukha). Artificial and synthetic fibres are worked widely, though again the area around Moscow remains important. In Siberia and in parts of European Russia textile industries in various forms have been developed in towns where there has been a deficiency of employment for women.

The over-all distribution pattern of industry is shown in Figure 7.1, which marks the location of industrial plants shown in Soviet atlases and economic geography textbooks and shows some interest-

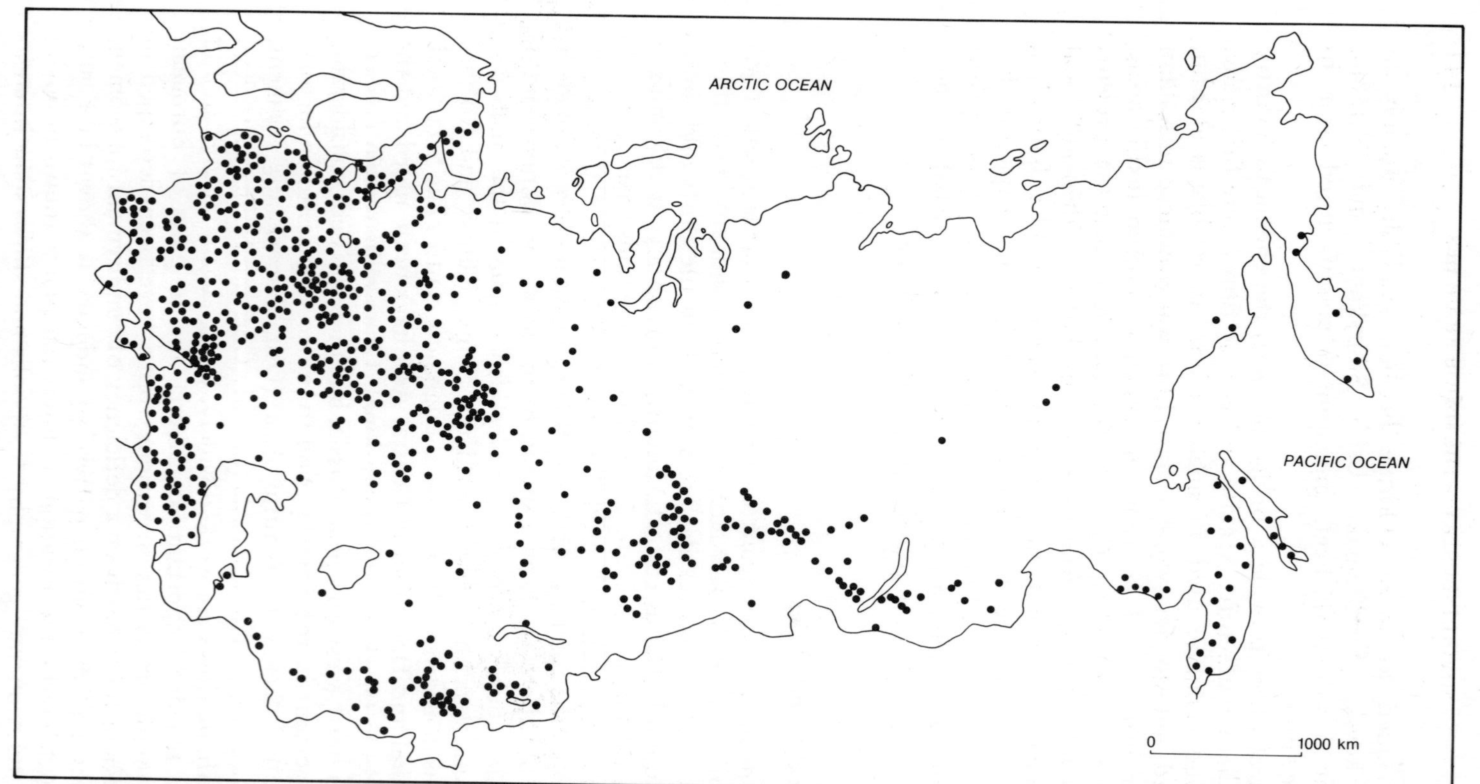

Each dot represents an industrial centre marked on major Soviet atlas maps of economic activities. The pattern displays the marked importance of European Russia, the Ural region and Western Siberia. Note also the smaller clusters in the Far East and Central Asia.

Figure 7.1 *The distribution of industry in the USSR*

ing correspondence with the map of population distribution (Figure 3.3). It stresses impressively the importance industrially of European Russia, notably the Ukraine and the central industrial region, besides the Ural and part of Western Siberia, with noteworthy clusters in Soviet Central Asia and in Transcaucasia. Compared with the distribution of industry before the Revolution, the spread on the contemporary map is wide, the product of over half a century of rigorous central planning directed by an amalgam of Marxist-Leninist political and economic theories, despite frequent revisions in their interpretation.

The pattern is the product of a maximisation of investment at the expense of consumption, while the plants themselves have been geared to manufacture capital equipment rather than articles for the Soviet citizen. The long-proclaimed aim of equalising industrial distribution among the regions has had perhaps less effect than might have been expected: in some respects, it seems to have been quietly overlooked. Much of this over-all pattern arose in the Stalinist period when political rather than economic considerations predominated in locational selection; this probably accounts for many of the remoter locations in Siberia, Central Asia and Transcaucasia. It should not be overlooked that the wartime 'scorched-earth' policy of moving plants from the path of the enemy into the interior also had a significant role. With the shift towards more sophisticated manufacturing industries and a greater involvement in world trade, besides more regard for economic elements in locational decisions, the concentration of industry in European Russia, the Ural and perhaps the more favoured parts of Western Siberia may expectedly intensify in the next decade. If appropriate data on the size of individual plants were, however, available, the effect of the gigantomania of the inter-war years and the more recent preoccupation with the idea of the territorial production complex would be reflected in a limited number of vast plants scattered across central and southern European Russia, the Ural and Western Siberia, overshadowing the broader scatter of the many small plants.

The total of Soviet industry is nevertheless impressive when we look at Table 7.1, extracted from the Soviet Statistical Yearbook. It claims a first or second place in world production and a first place in Europe for a wide range of products. (The reader may, however, like to reflect on the omissions from this list chosen by the Soviet state planners!) Their task in building this industrial structure has been far from easy: they have had to cope with the contradictions of the principles of Soviet planning, such as achievement of both regional specialisation and regional self-sufficiency, the equation of gigantomania in plants and the recent vast territorial production complexes with reducing transport demand to the minimum, while they have faced the serious contrasts in the geographical distribution of raw materials and energy sources with those of labour availability and consumer distribution.

Table 7.1 *Position of the USSR in world industrial production*

	World position	European position
Electrical energy	2	1
Petroleum (inc. gas condensate)	1	1
Natural gas	2	1
Coal	1	1
Pig iron	1	1
Steel	1	1
Coke	1	1
Chemical products	2	1
Mineral fertiliser	1	1
Sulphuric acid	2	1
Engineering products	2	1
Mainline railway locomotives	1	1
Tractors	1	1
Cement	1	1
Wool cloth	1	1

Source: *Narodnoye Khozyaystvo SSSR* (1978 edition).

7.6 Where to follow up this chapter

The problems of economic and industrial growth are examined in many texts, including Campbell, R. W., *The Soviet-type Economies* (Macmillan, London, 1974), Dyker, D. A., *The Soviet Economy* (Crosby-Lockwood, London, 1976), Gregory, P. R., *Soviet Economic Structure and Performance* (Harper & Row, New York, 1974), Kotkov, F., *The USSR Economy in 1972–1980* (Moscow, 1977).

'Economic geography' has been a major division along with 'physical geography' within the discipline in Soviet thought, covering virtually what in the West is called 'human geography', though the emphasis in the Soviet Union is strongly on industrial and agricultural aspects. An early and classic work is Balzak, S. S., Yasyutin, V. F. and Feigin, Ya. G., *Economic Geography of the USSR* (Macmillan, New York, 1949), while Baransky, N. N., *Economic Geography of the USSR* (Moscow, 1956), is a translation of a long-standing Soviet school textbook. Other major Soviet works include Breytermann, A. D., *Ekonomicheskaya Geografiya SSSR* (Leningrad, 1965), Cherdantsev, G. N. *et al.*, *Ekonomicheskaya Geografiya SSSR*, 3 vols (Moscow, 1956–8), Khrushchev, A. T., *Geografiya Promyshlennosti SSSR* (Moscow, 1969), Kovalskaya, N. Ya. and Khrushchev, A. T., *Ekonomicheskaya Geografiya SSSR* (Moscow, 1970), Lavrishchev, A., *Economic Geography of the USSR* (Moscow, 1969), Nikitin, N. P. *et al.*, *Ekonomicheskaya Geografiya SSSR* (Moscow, 1973), and Sushkin, Yu. G., *Ekonomicheskaya Geografiya Sovetskogo Soyuza*, 2 vols (Moscow, 1967–73).

Other useful sources are Dewdney, J. C., *The USSR*, Studies in Industrial Geography (Hutchinson, London, 1978), Meshcheryakov, V. *et al.*, *SEV – Printsipy, Problemy, Perspektivy* (Moscow, 1975), and Mathieson, R. S., *The Soviet Union: An Economic Geography* (Heinemann, London, 1975).

8

Transport – Holding the Soviet Economy Together

In the great distances of the Soviet Union it is not surprising that transport plays a vital role in holding the territory together. Transport cost in both capital and operating terms occupies a significant part in planning, and the problem of distance is a recurrent theme in the economic linkages within the Soviet economy, with every effort made to reduce the burden on a widely meshed transport network of mostly modest capacity. The transport infrastructure is a major consumer of national resources, perhaps absorbing as much as a quarter of all steel and fuel produced annually, while in 1978 9 per cent of the employed population worked in transport, a quarter of those workers on the railways alone.

Soviet sources stress the importance of transport as a unified system, in which each medium performs the tasks it can do best in contributing to a 'rational' distribution of productive forces. The concept of the unified transport system (Figure 8.1) has been increasingly emphasised in the 1960s as media diversification became more pronounced; before this, Soviet interest had been consistently focused upon the railways, though these still remain the prime haulier (Table 8.1). Now the view is to regard each medium as significant as a link in the transport chain, itself a link in the productive cycle of the over-all economy, though it is stressed that transport in itself is not 'productive' and therefore every effort must be to rationalise demand for it by effective central planning. In extolling the virtues of this view, Soviet sources point out that though all *wasteful* competition is eliminated, there may still remain room for 'constructive' inter-media competition. Every transport flow is to be

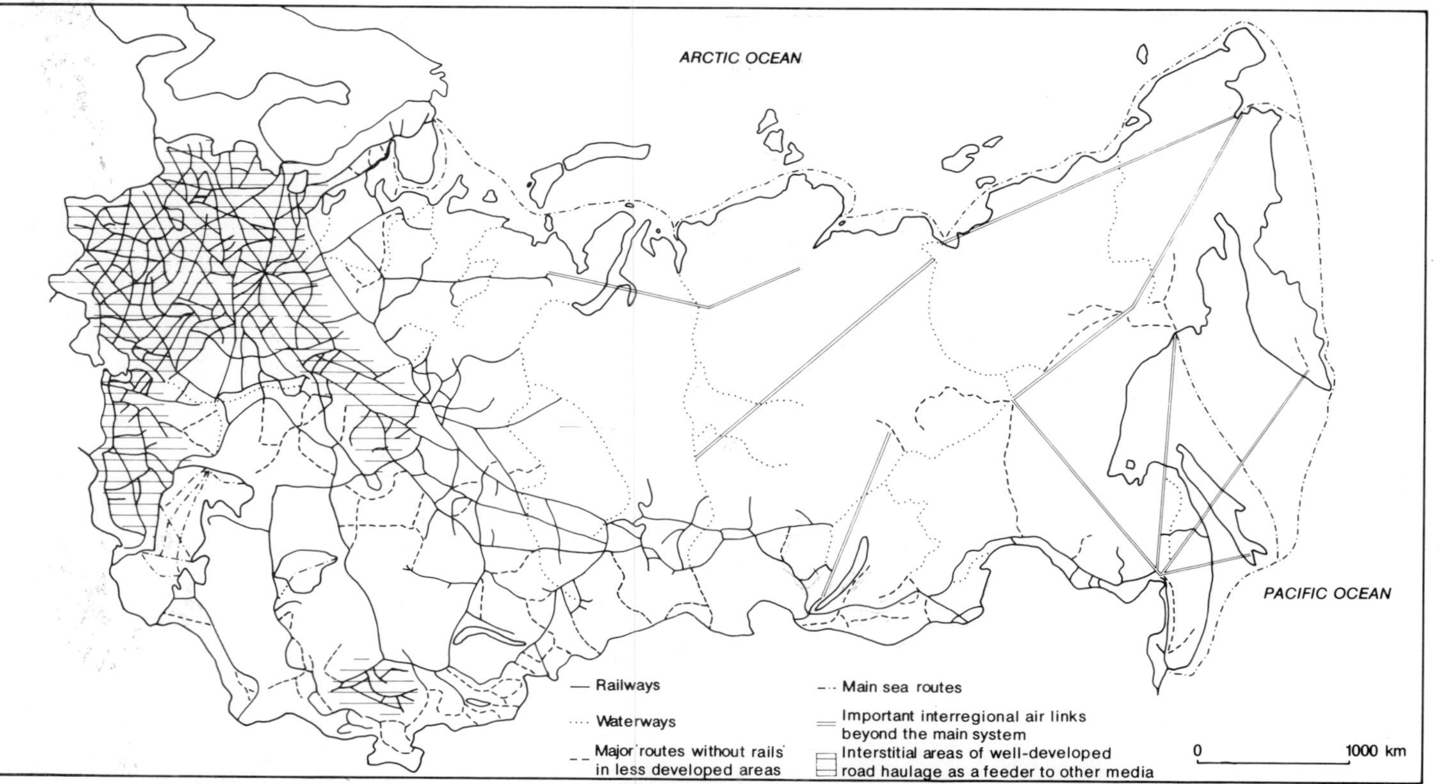

The map shows the composition of the major elements in the Soviet concept of a unified transport system.
Source: Kazanskiy, N. N. *et al.*, *Geografiya Putey Soobshcheniya* (Moscow, 1969).

Figure 8.1 *The Soviet concept of a 'unified transport system'*

Table 8.1 *Division of transport effort between the media*

	Total	Rail	Sea	Inland waterway	Pipeline	Road	Air
Freight traffic	*(milliard tonne/km)*	(%)	(%)	(%)	(%)	(%)	(%)
1940	494.4	85.1	5.0	7.3	0.8	1.8	Negligible
1950	713.3	84.4	5.6	6.5	0.7	2.8	0.01
1978	5,947.9	57.7	13.9	4.1	17.6	6.6	0.1
Passenger traffic	*(milliard passenger/km)*						
1940	108.7	92.4	0.8	3.5	—	3.1	0.2
1950	98.3	89.6	1.2	2.7	—	5.3	1.2
1978	841.8	39.5	0.3	0.7	—	42.9	16.6

Source: *Narodnoye Khozyaystvo SSSR* (1978 edition).

regarded as one operation, with the media mix to achieve it governed by factors such as volume, distance, frequency of movement and orientation, with special attention given to terminal facility requirements. All movements should be classified as 'local', 'interregional' or 'international', and the media and routes selected accordingly.

Transport infrastructure has, however, been much influenced by policy decisions dating back to the 1920s, when basic Soviet planning concepts were being formulated. These were implemented by *Gosplan* in the inter-war and immediate post-war years, being modified only in the 1960s, but possibly technological developments in the remaining years of this century will further accelerate change. The search for a high degree of self-sufficiency produced strongly introvert tendencies in transport, turning it towards interior traffic-generating districts rather than to ports or border crossings, with little interest shown by the planners in designing a system as part of the larger world pattern. Consequently the Soviet transport structure became strikingly 'continental', with distorted haulage distances much longer than those commonly regarded as optimal, reflected notably in excessively long hauls by railway or inland waterways of low-grade bulky commodities.

Although strategically and politically desirable, the dispersal of industry through the regions during the inter-war years posed critical transport problems, most serious in the eastern regions where the transport infrastructure was distinctly backward. One attraction (at least in theory) of the regional dispersion of industry and local self-sufficiency was an economy in transport effort by moving industry to its raw-material and fuel bases, while there were hopes of cutting out wasteful cross-hauls. There are, of course, limits to regional autarky and the elimination of the demand for transport, for there is an uneven spatial distribution of raw materials and fuel resources. In practice, dispersal of industry had its limits, partly because of production-planning reasons but also because of the constraints imposed by the inadequacies of the transport infrastructure. Too high a level of dispersion would not only have placed an insuperable demand on transport but would also have made it impossible to achieve reasonable economies of scale through building and operating plants of requisite size, and the resulting pattern became one of a limited number of major industrial nodes linked together by a trunk transport system. The territorial expression came in the gigantomania of the 1930s and more recently in development of the concept of the territorial production complex.

The development of the Soviet economy saw a considerable re-orientation in the transport system from tsarist times, when it had been geared to an essentially agricultural economy. The main movements then were of food from surplus areas to those with a deficit or to the ports for export, and traffic was light and infrequent. The industrialisation set afoot after the Revolution demanded not only new directions in which to carry raw materials to the new factories and to distribute their products, but it also demanded frequent and heavy traffic, in effect a 'moving belt' rather than the previous seasonal shifts. The Soviet transport system has a striking dimensional form, extending some 4,500km from north to south and some 9,000km from east to west, with the pre-eminent direction of movement along the east–west axis, which puts the main natural waterways, with their predominantly north–south orientation, at an immediate disadvantage. With immense tracts of virtually uninhabited territory and, as noted already, with over two-thirds of the population living on one-sixth of the area, west of the Volga and south of Leningrad, where over half the national investment is made, this area claims much the densest transport mesh, from which long branches stretch out into the near empty periphery to the few but vital growth areas.

The physical environment of the Soviet Union (see Chapter 1) has exercised a considerable influence over the choice, use and distribution of the different transport media, even without pressing geographical determinism imprudently far. The anomalous cold – the long frozen season and the violent blizzards – is of particular significance, while in the south aridity, compared with wetness in the north, also plays a role in influencing the nature of transport links. For the civil engineering structures needed by the transport media, permafrost over almost half the country is an additional hazard.

In terms of traffic effort for both goods and passengers, the railway remains the principal haulier. Road haulage in terms of goods traffic is less significant, though it does perform an appreciable proportion of passenger journeys. In terms of total originating tonnage and total originating passengers, road haulage shows a higher proportion than railways, but it is, however, the shortness of the average road haul for both that reduces substantially the roads' share in traffic terms (Table 8.2). Inland waterways contribute an extremely modest share of goods traffic and an even smaller share of passenger traffic. There has been a striking rise in the volume of traffic in a closely defined range of commodities handled by pipelines. The generally long hauls of freight by sea-going ships are reflected in the exceptionally modest share of originating tonnage but an appreciable share of total freight-traffic effort. Sea-going passenger traffic on all counts is negligible, mostly restricted to internal movements on the Black Sea, Caspian Sea and in Far Eastern

Table 8.2 *Division of originating goods and passengers among the transport media*

	Total	Rail	Sea	Inland waterway	Pipeline	Road	Air
Goods	(million tonnes)	(%)	(%)	(%)	(%)	(%)	(%)
1940	1,578.5	38.3	2.1	4.7	0.5	54.4	Negligible
1950	2,834.5	29.4	1.2	3.2	0.5	65.6	0.01 (est.)
1978	28,243.5	13.4	0.8	1.9	2.1	81.8	0.01 (est.)
Passengers	(millions)						
1940	2,050.5	67.2	0.5	3.6	—	28.8	0.01
1950	2,279.8	51.0	0.4	2.4	—	46.2	0.06
1978	44,271.5	8.2	0.1	0.3	—	91.2	0.2

Source: as for Table 8.1.

waters, though tourist cruising is increasing. Airways hold a growing share of passenger traffic and a rapidly rising number of originating passengers. Their contribution, on the other hand, to freight movement is small, comprising shipments of high-value goods, mail and newspaper plates set in Moscow for printing in the provinces.

It is interesting to note that for railways and inland waterways, hauls are much longer on average for goods than for passengers. Railway average distance for passengers is reduced by the component of commuter journeys, while short pleasure trips by boat or the use of ferries have the same effect for inland waterways. Over 65 per cent of all goods sent by rail move over 200km, while distances over 500km account for 44 per cent of all goods. It is expressed policy to transfer all hauls under 80km from railway to road transport, though nevertheless 20 per cent of all goods sent by rail still move less than 100km. On the other hand, there is considerable development of road haulage in central European Russia and the Ukraine for distances up to 200km.

8.1 **Railways** (see Figure 8.2)

The importance of Soviet railways is impressive when it is noted that they handle 45 per cent of world railway freight traffic and that average freight density per kilometre of route is five times that in the USA, while the Soviet authorities claim that the density of passengers per kilometre of track is also several times the world average. Considering, however, the continental dimensions of the USSR, such an important role for railways is perhaps not unexpected, for both the distance factor and the nature of traffic as well as environmental conditions are peculiarly well suited to rail haulage. The railways have benefited from a conscious decision of the 1920s not to encourage road traffic, for at that time it was firmly believed that substantial economies in investment and resources would be achieved by developing an existing medium rather than an embryonic one. Long after the reality of the initial decision had faded, the powerful railway lobby perpetuated railway dominance. The continuing importance of the railway is reflected in the on-going extension of the network through several major projects. Roads and inland waterways are still regarded essentially as feeders to the railways, while pipelines and airways are seen as relieving some of the railways' burdens in certain critical areas of freight and passengers.

In general, railway construction has been favoured by an absence of major relief obstacles, though there are features of the microrelief that do pose formidable civil engineering problems. The Ural ranges across the main east–west routes do not, however, constitute

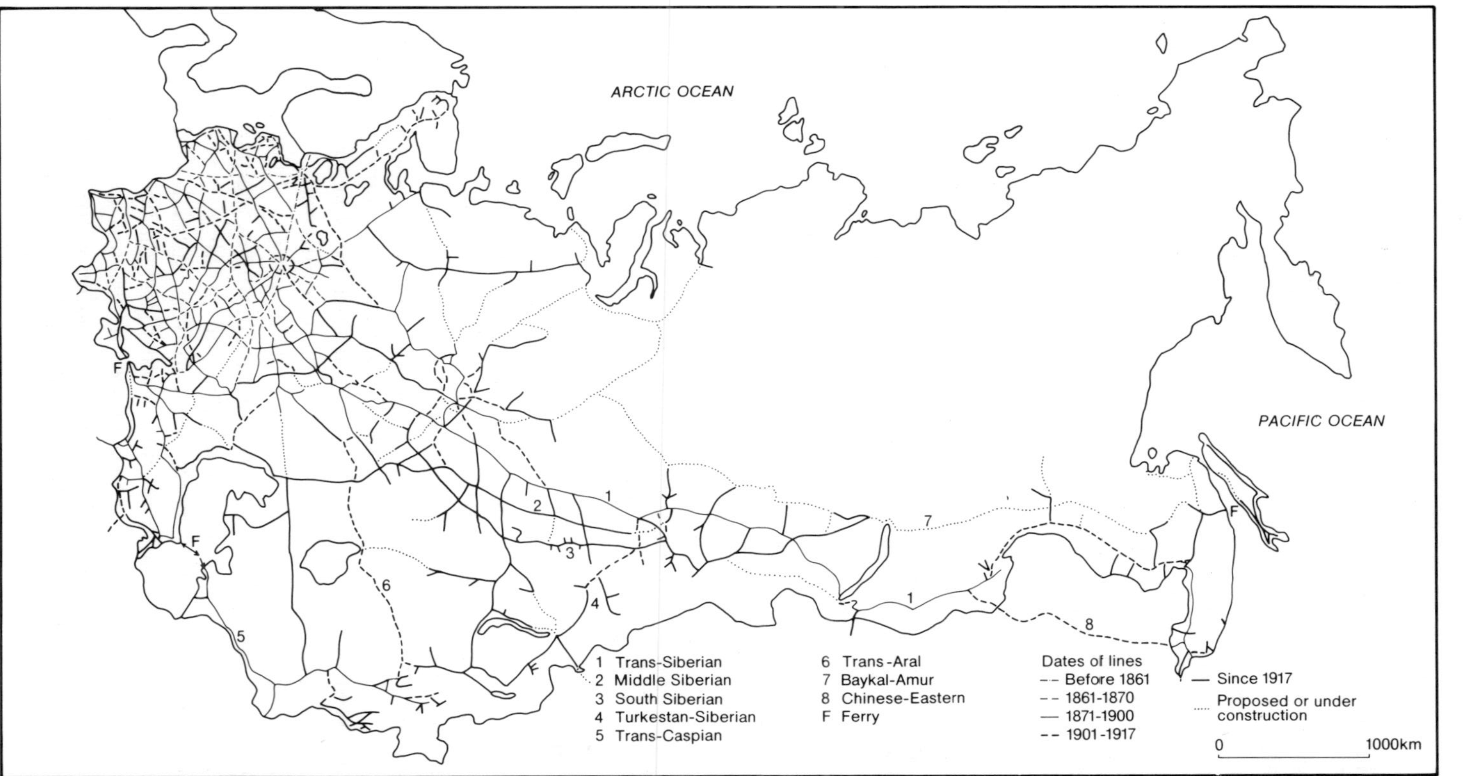

The map of dates of railway construction reflects the major development in the eastern regions since 1917. The Baykal–Amur project should be complete about 1983.

Sources: various Soviet railway atlases.

Figure 8.2 *The Soviet railway system*

a serious obstacle to communication between 'European' and 'Asiatic' Russia, because their central section is crossed by several easy passage-ways. In southern Siberia, although relief is more varied, comparatively easy routes again exist through the rough dissected country of Baykalia. In the Caucasian isthmus, a littoral shelf provides a convenient way to circumvent the formidable ranges of the Great Caucasus, while a central depression between these ranges and the Little Caucasus to the south provides an east–west route, though it has a difficult section through the narrow Suram Pass. The few railways that penetrate into the mountainous border country of Central Asia do so by using obvious valley routes.

The predominance of flat to undulating topography is reflected in the fact that some three-quarters of the total route length is straight track and has gradients gentler than 1:166, with one one-fifth of the route having gradients steeper than 1:60. Micro-features that influence the trajectory of railway routes include extreme gullying in steppelands, the extensive swamp and marsh, and even the wide rivers with their broad flood plain meadows regularly inundated every spring. In arid areas drifting sand is a problem, though over the great plains drifting snow in winter can be a serious hazard. A problem of the great plains, with their deep mantle of soft deposits, is that good railway ballast is hard to find. This is important because, with waterlogging in spring over large tracts and danger of frost-heaving in winter, good drainage of the road bed is vital. Original economies in building costs, resulting in sharp curves and undulating track, sometimes with short sharp gradients, imposed operating restrictions on axle loads and speeds which have been real constraints to increasing the capacity of many lines without a massive capital investment programme. If it is considered that one kilometre of single track requires an input of 100–170 tonnes of steel rails, 185m^3 of wooden sleepers and 1,500m^3 of scarce ballast, the impact of distance on national resources becomes apparent, especially when the construction of a branchline like the 720km route from Tayshet to Ust-Kut on the Lena is considered.

Railway operations also suffer in the great winter cold, when fuel consumption is raised. Heat loss on steam locomotives was particularly serious, while coal froze in the tender and problems existed with water supply. Even with diesel locomotives, special fuel supply systems are needed because in great cold diesel oil becomes highly viscose. Bulky wet loads like ores freeze in wagons and unloading becomes difficult, and there is constant trouble with frozen points and ice on overhead catenary. Winter cold in the north and aridity in the south long posed a problem of water supply for steam locomotives: one attempted solution was use of the condensing tender, but this adds considerable weight. The open plains are frequently swept by high winds, and a costly solution to the hazard is provision of long shelter belts, of which there are well over 37,000km. Permafrost is a difficulty in parts of Siberia, with a real danger to civil engineering structures through damage by frost distortion wherever the thermal balance is disturbed, and every spring long lengths of track have to be virtually relaid. Operations are everywhere upset by temporary speed restrictions and axle-load limits.

Railway building began in 1837 with the short St Petersburg–Tsarskoye Selo line, soon followed by politically important lines like St Petersburg–Moscow (1843–51) and the St Petersburg–Warsaw railway (1848–61). It was not until the 1860s that railway construction began to accelerate, first to carry grain to Baltic and Black Sea ports for export and later to join Moscow to nearby towns like Gorkiy and to those districts that supplied its food. Strategic reasons motivated railway construction in Caucasia in the late 1870s, but progress was slow, for Russian engineers for the first time met difficult terrain. This period also saw the building of lines to serve emerging industries in areas like the Ukrainian Donbass. By the mid-1880s building across the Volga to the Ural towns was under way, and in 1892 work on the Trans-Siberian Railway began. By 1898 Irkutsk had been reached and by 1905 the train ferry across Lake Baykal had been replaced by a difficult 70km section with thirty-eight tunnels around the southern shore. Travel to Vladivostok was possible across the Chinese Eastern Railway through Harbin in Manchuria, but with the loss of the war with Japan an all-Russian route had to be built. This was completed between 1908 and 1916 along the Amur via Kuenga and Khabarovsk, where it joined the Ussuri railway to Vladivostok, built in 1891–9. The 1880s had also seen building of the Trans-Caspian Railway and in 1905 the Trans-Aral Railway linked it to the Volga. In 1915 two strategically important railways, those to Murmansk and to Arkhangelsk, were opened. By 1917 some 70,000km of route, albeit varying much in quality, formed a network that was neither rudimentary nor incomplete for contemporary requirements.

Following the rehabilitation of the railways after the Revolution and civil war, the first task was to develop major interregional links, creating a widening gulf in standards between main and secondary routes. An early project was completion of a third trunk route from the Moscow district to the Donbass, created by linking together and upgrading existing secondary lines, while other mostly strategic lines begun before 1917 were quickly completed. Several major projects needed by new economic developments were also begun, including the strengthening of the link between the Ural and Kuzbass for the planned *Kombinat*, better rail links from the Volga to the Ural, and the provision of rail access to Karaganda. The greatest achievement in the early 1930s was completion of the Turksib Railway, joining Western Siberia to Central Asia, a project planned before the Revolution. In the later 1930s interest shifted to providing more and better interregional links, notably between the Caucasian lands, the Ukraine and the Volga, where a railway parallel to the river was begun, for north–south links in the basin were weak. There was also the duplication of routes in Western Siberia, and in 1938 a vast plan was unveiled to duplicate the Trans-Siberian between Lake Baykal and the Pacific shore across exceedingly difficult terrain. This was basically a strategic line in case the Japanese in Manchuria were to cut the Amur section of the Trans-Siberian Railway where it ran close to the border, but further news of its progress was not given. Wartime building concentrated on offsetting and bypassing territorial losses to the invading Germans. Most important was a long branch-line to the Pechora coalfield, built regardless of cost when the Germans captured the Donbass. Another major line was begun around the northern Caspian shore to join Caucasia to Central Asia and the Volga, while work went ahead on the Volga parallel line, now strategically of a high order. Access to the all-weather port of Murmansk was improved by a line giving a connection with Arkhangelsk, again bypassing territory lost to the Germans.

Restoration of the railways took the first post-war decade, though some lines in western districts were not rebuilt, particularly narrow-gauge routes in the Baltic republics and Byelorussia. Several new lines to serve the Ural–Volga oilfields appeared, and the general emphasis lay between the southern Ural, Kazakhstan and Western Siberia. Karaganda was joined to Central Asia by a line along the western shore of Lake Balkhash, while Central Asia was joined to the Volga by a line along the Amu-Darya and then across the barren Ust-Urt Plateau, with a major branch to the Mangyshlak oilfield. Several long branches were also built into the west Siberian lowland to serve new oilfields, including a very difficult one to the Urengoye field in the Taz basin. An important link in southern Siberia has been provided by the line from the Kuzbass via Abakan to Tayshet on the Trans-Siberian line. In Siberia a major early post-war completion was the line to Ulan Bator, capital of Mongolia, later extended to Peking, while another line was completed to the Dzungarian Gates in Kazakhstan, originally part of a long Friendship Railway to join the Chinese railways in Sinkiang, but the Chinese have not yet completed their section westwards from Urumchi.

In 1946 Soviet sources revealed the Komsomolsk–Soviet Harbour section of the Baykal–Amur project as complete (train ferry across the Amur) and in 1954 the western section from Tayshet to Bratsk and the Lena was shown to be in use. The tenth Five-Year Plan (1976–80) reactivated the Baykal–Amur trunk project on a modified alignment, to help relieve pressure on the main Trans-Siberian line but also to accelerate development of the Aldan and Vitim areas. It would appear from evidence now available that work had in fact been continuing since the early 1950s on some of the main civil engineering features like tunnels. In association with this project, a bridge has been built across the Amur at Komsomolsk and a train ferry links Soviet Harbour to the now much extended and modernised railways of Sakhalin. A line associated with the Baykal–Amur project has also been built northwards from the Trans-Siberian to open up the coal deposits of southern Yakutia. There appears to have been a net gain of 25,000km of route since 1946, excluding a considerable length of industrial railway.

At various dates some remarkable railways have been proposed. The idea of a line to Norilsk now looks as though it might materialise, for the railway has reached Urengoye not far from the Taz river and an extension east to the Yenisey is quite possible. Work on such a line from Salekhard on the lower Ob to Norilsk was begun shortly before Stalin's death but shortly thereafter abandoned. On the other hand, a railway proposed in the early post-war years from the lower Volga to Stavropol and then directly across the Great Caucasus ranges to Tbilisi is less likely to be undertaken. Equally remarkable is the proposal for a railway north along the Okhotsk coast to Anadyr on the Bering Strait's approaches, first suggested by American interests around the turn of the century. Another unlikely line suggested by some Soviet maps is north to Yakutsk and then along the Lena to Tiksi on its delta.

Illustrative of the problems of railway operation in the Soviet setting is the question of motive power. Although the political pressure for electrification throughout the country stimulated by Lenin brought plans for railway electrification, generating electricity was slow to develop and priority was diverted to 'productive' industry. Consequently in the 1930s only suburban commuter lines around Moscow and Baku and such heavily graded freight lines as the Suram Pass route in Georgia and the Kizel–Sverdlovsk line in the Ural were electrified. Experiments at this time were conducted with diesel traction, notably in Central Asia where water for steam locomotives was short, but after the mid-1930s interest shifted to steam locomotives with condensing tenders. The steam locomotive was, in the prevailing conditions, most attractive, because fuel was plentiful, it was easy and cheap to build, simple to maintain and operationally reliable, while labour costs were a secondary consideration. The real problem arose with the growth of traffic, because increasing frequency of trains meant costly double-tracking and more and longer passing places on both double and single track. Heavier trains to offset something of the increase in frequency would have needed steam locomotives with more power but also higher axle loads that demanded much strengthened track. Complete rebuilding of the track would have been an impossible task in terms of time and cost over the distances involved. The alternative to heavier axle loads for locomotives was either excessively long rigid-frame locomotives, with consequent damage to track and points, or articulated locomotives, too complex and awkward to maintain in the Soviet setting.

Both diesel and electric traction, despite higher capital costs, give much better power–weight ratios, so that heavier trains can be handled faster without any appreciable increase in axle loads. The extra cost of locomotives and lineside equipment is offset by the saving in not having to rebuild the track and roadbed. By the 1950s both rising oil output and increasing electricity generation made the use of either form of traction possible. Electrification was chosen for the mainlines, using the French 25kV system for much of the route because of its savings in scarce copper by its light catenary, its use of industrial current dispensing with lineside sub-stations, and its light overhead structures, so that the route length has risen from 3,000km in 1950 to 41,000km in 1978. The electrified route and that still planned for inclusion shows a remarkable similarity to the main T-shaped arterial railway system. Diesel traction is now widely used on secondary routes.

Before the Revolution train services were of infrequent, light trains, but with the improvement of track and equipment in Soviet times freight services now combine the frequency of European railways with train weights comparable with American practice. Like all railways, traffic is unevenly spread across the system; according to Soviet sources, 86 per cent of all freight traffic is concentrated on 46 per cent of the route length, while of the total the quarter of the route which is double-track claims 64 per cent of the traffic. A further element of concentration is reflected by the fact that 80 per cent of all wagons loaded daily originate in six railway directorates – Donetsk, Tomsk, Pechora, Karaganda, Sverdlovsk and Kirov. Goods like iron and steel semis, wood and grain have average hauls over 1,000km; the average haul of coal (692km) is over five times longer than the corresponding movement in the United Kingdom. Cotton bales on average move 3,327km and salt fish 2,271km. With such long hauls, empty running of wagons becomes a serious problem and unavoidable empty running annually equals wagon capacity for 130 million tonnes of goods. Refrigerated wagons do 47 per cent of their running empty, open wagons about 26 per cent, and the over-all average is 38 per cent. Reduction of empty running is all the more important with high-capacity wagons that have enabled a 25 per cent increase in train weight to be achieved without longer trains or heavier axle loads.

Apart from the busiest trunk routes where several passenger trains may run daily, most lines have only one or two passenger trains each way daily. With the extremely long distances of some trains, even to provide one train a day in each direction may require several train sets when the journey itself takes several days from end to end.

Evidence from commodity and passenger flow maps for the railways published in Soviet textbooks during the 1960s illustrated the concentration of the larger part of the total traffic, as already noted, on a small part of the total route. In spatial terms this produces a T-shaped pattern of a limited number of super-trunk routes into which traffic is fed from the rest of the system. The head of the T is formed by the north–south trunk routes linking the European north-west via the central industrial region to the industrial south (the industrial towns of the eastern Ukraine). The main stalk of the T comprises the few lines that run from these north–south trunk routes eastwards across the Volga and Ural into Siberia. As traffic densities fall appreciably east of Baykal, this may be regarded as the foot of the T. The rapid growth in containerised traffic by rail from Japan and Eastern Asia to Europe has been putting an increasing strain on the capacity of the east–west routes across Siberia.

8.2 Inland waterways

Inland waterways have provided the traditional means of transport in Russia, but despite substantial increases in traffic volume, their share of the national total has fallen strikingly. Much of the problem arises because the rivers flow either northwards or southwards, whereas contemporary traffic demand is principally for east–west movement. Another difficulty is their limited period of navigation, either because of the winter freeze or summer low water. Movement from one system to another involves slow and costly transhipment, while to build canals to join together rivers is exceptionally expensive. Of 3 million km of river, a little over 140,000km are suitable for shipping, of which 82,000km have guaranteed depths. Artificially improved waterways and canals comprise about 20,000km of route and handle 60 per cent of all waterborne traffic, with the Volga system, serving a vital north–south axis in European Russia, carrying over half the total waterborne freight traffic. The rivers handle about 520 million tonnes of goods a year, of which over 300 million tonnes are mineral building materials and some 67 million tonnes are lumber and firewood, with some 25 million tonnes of coal and coke and 20 million tonnes of petroleum and petroleum products. Exhortations to industry to make more use of waterways receive little response, since in Soviet economic conditions it is difficult to stockpile for the closed season and consequently preference remains for the 'moving belt' that the railways can provide.

As noted, the most intensively used waterway is the Volga with its tributaries like the Kama, the Oka and the Belaya. Within the Volga drainage basin live well over 70 million people. The importance of the Volga waterway system is shown by the number of canals that link it to other waterways. In the north the Volga–Baltic Canal (1964) allows 5,000-ton vessels to enter the river from the Baltic, while in the south the Volga–Don Canal (1952) corrects one of nature's errors by giving access across the narrow interfluve between the two rivers, so allowing vessels to reach the Sea of Azov and the Black Sea, and Moscow is joined to the Volga by a canal more direct than the natural but meandering route along the Oka valley. An artificial channel into deep water also gives shipping safe passage from the Volga into the Caspian Sea at the delta. An early nineteenth-century canal, now virtually unused, allows small vessels access from the Volga into the basins of the Northern Dvina and Sukhona.

Volga navigation illustrates the problem faced on most Soviet rivers. In late summer navigation on several reaches is limited by low water; in spring meltwater raises the level and floods spread across the low-lying banks, so that with the current much accelerated and ice-floes swirling downstream, navigation is again hazardous; and the long frozen period of winter stops boats using the river. On the upper reaches navigation is possible for six to seven months each year and up to eight months on the lower reaches. Regulation of the channel has been carried out on many sections, both by dredging and by marking it with buoys. It was planned that the cascade of hydro-electric barrages would help navigation, but these have not been as helpful as anticipated. The large expanse of shallow water dammed behind the barrages turns very rough in certain wind conditions, so that the traditional shallow-draught, flat-bottomed river boats become unseaworthy. By slowing the flow of the river, the barrages have led to a longer period of icing, up to ten days in some places. At least, by controlling the flow, summer low water can be shortened and even eliminated in certain sections, though it has become apparent that silting has been quicker than expected and dredging more necessary in some reaches.

The second most important river is the Dnepr, the use of which was greatly increased by completion in 1932 of the Dneproges dam that made possible navigation of the former 80km of rapids between Dnepropetrovsk and Zaporozhye, while in the early 1950s navigation on the lowest reaches was eased by completion of the Kakhovka barrage. A canal of limited capacity joins the upper Dnepr to the Bug and another to the upper Western Dvina, but there have been grandiose plans to construct high-capacity links to the Vistula and central European rivers, so providing an immense internal waterway ring. With a navigable season of 265 days at Kiev, problems are similar to the Volga, with a long winter freeze and summer low water. Like the Volga, shipping is both by large train and self-propelled barge, while modern pushtug and barge systems are now being increasingly employed.

Most rivers of the Baltic and northern Russia have only a limited use because of their size and profile, while ice becomes an increasing hindrance to the north. Mostly they are used for floating timber, notably the Sukhona, Vychegda and the Northern Dvina. Between the wars a canal was built from Lake Ladoga via the Karelian lakes to give access from the Baltic to the White Sea; principally used for timber movements, it also carries fishing boats and small naval ships.

The great Siberian rivers have been little developed for transport and are in several respects difficult to use. They all suffer from short navigation periods – from 150–200 days in their southern reaches to less than ninety days in the northern reaches. Thawing first in the upper, southern reaches, masses of water pour down on to the still frozen lower, northern reaches, causing immense floods, while massive ice-floes jam together in dams that pond back water until they break, when catastrophic torrents pour northwards. In summer the warm water flowing north mixing with cool arctic air produces thick fog for long periods in the lower reaches. Rapids block the course of the Angara, a useful link from the Yenisey to Lake Baykal, but a cascade of hydro-electric barrages when complete will open it to navigation. On the Yenisey sea-going ships can sail up to Igarka, over 700km from the mouth and the point where the current of the lower river makes the handling of barge trains difficult. The lower Lena is little used, scourged by ice in summer, inadequately charted and with an awkward delta, closed by jammed ice-floes in many summers. Development of oil and gas resources in the Ob–Irtysh basin will undoubtedly accelerate their rivers' use. Timber is a main freight commodity, moving both upstream to the railway ports along the Trans-Siberian railway and downstream for export, but on the Ob–Irtysh system petroleum and coal is carried upstream, whereas on the Yenisey ore and metal from Norilsk also goes upstream. The Lena is used mostly for traffic between the Ust-Kut railhead and Yakutsk.

In the Far East the Amur is navigated upstream to Sretensk on its tributary, the Shilka. Ice-free for 150–90 days, the river has an acute summer low water, but there are also dangerous flash floods. Its estuary is almost unnavigable and the river is generally not much used below Komsomolsk. The Ussuri is also relatively unimportant. In Central Asia some navigation takes place on the Amu Darya and to a lesser extent on the Syr Darya. Cotton, wool and some ore are sent downstream and across the Aral Sea to Aralsk for onward shipping by rail. Return freight is mostly grain and some manufactured goods. Although ice is unimportant, in summer shallows limit navigation.

8.3 Sea-going shipping

So long as the emphasis was on a closed economy, the small Soviet mercantile marine comprised mostly vessels for the limited coastal traffic. Sea transport between the ports of European Russia and those of the Far Eastern territories faces a long and circuitous journey, while the shorter northern sea route along the Soviet Arctic coast is only open for a few summer months. The Baltic and the Black Sea (with the Sea of Azov) have their entrances covered by foreign powers, but their ports, particularly those of the Baltic and the northern shore of the Black Sea, suffer from varying degrees of icing, and the Sea of Azov can be cut off in winter by the freezing of the Strait of Kerch. The inland Caspian Sea has conditions that are truly 'sea-going'. The Soviet Far East coast also suffers long periods of ice in winter. On the Arctic coast good anchorages are rather few, and Murmansk, most important of the northern ports, is surprisingly kept open through the winter by a warm branch of the North Atlantic Drift. Several ice-free outports have been developed since the 1950s to augment its capacity. The difficulties of keeping ports open in winter and the risks to navigation in icy waters have been seen by many students as a key to the political geography of Russia since tsarist times, mirrored by the search to secure 'warm-water' ports.

The opening of the Soviet economy to a greater participation in world trade since the 1960s has been reflected in the growth in volume of goods shipped by sea from a mere 33 million tonnes in 1940 and 54 million tonnes in 1955 to over 229 million tonnes in 1978. At the same time, a substantial growth in the Soviet merchant fleet has taken place, largely by the purchase of vessels from other Comecon members and by building in Soviet yards. There has also been an agreement in Comecon that at least half socialist *bloc* trade should be carried in ships belonging to it, while the Soviet Union has sought to invade the world charter market and the cruise trade. Traffic between Soviet ports on the same sea dominates by far, while in second place is international trade, with traffic between Soviet ports on different seas a poor third. By far the busiest ports are those of the Black Sea and the Sea of Azov, while Caspian ports are next in order, followed by those in Far Eastern waters, Baltic ports and the northern seas (chiefly Murmansk and Arkhangelsk).

Navigation in the Black Sea and the Sea of Azov is reasonably easy, though care is needed in their northern waters because of extensive shallows. Off the Crimean and Caucasian coasts wind and currents can be tricky, while spring sea mist and summer dust-storms can make visibility extremely poor. Coal, ores and petroleum are exports in international trade, with imports comprising foodstuffs, rubber, natural fibres and manufactured goods. Odessa is the principal port, while Nikolayev, 90km from the sea up the estuary of the Southern Bug, is the main shipbuilding centre. Access to the Sea of Azov is through the difficult Strait of Kerch, kept open by icebreakers in some winters. Zhdanov exports coal and metal against imports of ore, while Rostov, some 50km up the Don, is a large general port, though the winding channel is hard to navigate. On the Caucasian Black Sea coast, Novorossiysk is the main port, shipping out cement and petroleum, also sent from Tuapse and Batumi.

Icing conditions are more severe in the Baltic, especially in the Gulf of Finland and Riga Bay. There is less coastwise traffic than in the Black Sea, though Leningrad is the most important of all Soviet ports, handling a wide range of cargoes, notably imports. The other ports are of modest capacity. The former German ports of Kaliningrad and Baltiysk are mainly used for naval purposes, but an oil terminal has been built at Ventspils.

The long coast of the Far East accounts for only a small part of Soviet seaborne traffic. All the ports except Vladivostok and Nakhodka suffer badly from winter ice, with ports on the Okhotsk Sea closed for more than 200 days. Vladivostok, the main port, is nowadays used only by Soviet ships, and a vigorous international traffic and a ferry service to Japan is run from the new port of Nakhodka, ice-free for longer periods than Vladivostok, while additional capacity is being developed further east along the coast at Vostochnyy. Development of Soviet Harbour and Vanino has included a train ferry to Sakhalin and will expand when the Baykal–Amur trunkline is completed. Nikolayevsk at the Amur estuary remains useful only for small vessels. On the Okhotsk Sea, Magadan (founded 1931) is the main port, shipping ores from the Kolyma basin, and kept open as long as possible by icebreaker, replacing the old port of Okhotsk. In Kamchatka, Petropavlovsk, seldom icebound and limited to vessels up to 5,000 tons, serves as the winter harbour of many vessels from the northern sea route.

The difficulty of maintaining contact between western and eastern ports in the Soviet Union encouraged development of an 'all-red' strategic route across the long and hazardous peripheral waters of the Arctic Ocean. First navigated successfully in 1878–9, a special Soviet administration in Leningrad was established to develop its use, with which the famous expedition by the icebreaker *Chelyushkin* was linked. It has been suggested that some inter-war success in opening this route must be ascribed to improvement in the general Arctic climatic situation, though undoubtedly improved techniques and equipment played a part. Year-to-year conditions vary considerably and some students suggest the recent mild period is now over and deterioration of the navigation season may be expected. Certainly, since the Second World War improved weather forecasting has been important, just as has the use of atomic icebreakers since the late 1950s.

Murmansk and its outports (mostly developed since the 1950s) at the western end are among the few Soviet ports ice-free in winter, but Arkhangelsk is closed mostly from November to May, though it may be kept open in good years by icebreakers. The White Sea, with its narrow entrance, is closed throughout the winter. In May of each year access may be had to the Kara Sea, by difficult navigation through its narrow entrances. The main ports to which ships go are Novyy Port or Labytnangi on the Ob, and Dikson, Dudinka or more usually Igarka on the Yenisey. Navigation further east through the Vilkitskiy Strait is always hard because of mist and grounded icebergs. Most ships that go through this strait call at Norvik on the Khatanga estuary, while Tiksi, the roadstead for the Lena delta, is a common port of call. The easternmost ports are Nizhne Kolymsk, Ambarchik or Pevek on Chaun Bay, though these are usually serviced by vessels from Far Eastern ports. The first port on the Pacific, Provideniya, was important for handling Lend Lease supplies during the Second World War. The Soviet authorities will now pilot foreign vessels through the northern sea route for appropriate payments, but there seem to have been few takers.

The immense Caspian Sea is in many respects an extension of the Volga waterway, but navigational conditions are more akin to the Baltic or Black Sea. Baku is the main port in the south and Makhachkala in the north, both maintaining ferry connections (some vessels are over 5,000 tons) with the eastern shore, notably Krasnovodsk. A lot of traffic in the north passes by an artificial channel through the delta to the Volga port of Astrakhan, while most recently new oilfields in the Mangyshlak Peninsula have made Shevchenko a potentially important port. Bekdash on the Kara Bogaz Gol exports salts produced in that immense natural evaporating pan.

A critical difficulty is that few Soviet ports can take the largest modern bulk carriers and tankers fully laden, and even some that could do so are debarred because their approach waters are too shallow or otherwise limited. Shallow draught vessels of about 100,000 tons can get into the Baltic with difficulty, while their entrance to the Black Sea is likewise extremely hazardous. Ports capable of taking 100,000-tonners on the Black Sea appear to be Odessa, Ilichevsk and Novorossiysk as well as Batumi and the offshore roadstead at Poti; in the Baltic, Riga roadstead can take them and possibly the Ventspils tanker terminal when complete. In the north, Murmansk seems able to handle big vessels with a good deep water approach. In the Far East, Nakhodka seems capable of handling such large vessels, though it is claimed that the 'largest vessels afloat' can be taken in the outer harbour at Vladivostok and the new port of Vostochnyy (when finished) will be able to handle such vessels.

8.4 Road transport

Road transport in the Soviet Union is primarily concerned with short-haul distribution in both the towns and the countryside, where it acts notably as a feeder to the railways or river wharves. It is generally regarded as a haulier up to 80km, though some Soviet sources suggest its role is in 'door-to-door' haulage of up to 200km, and certainly some road haulage and inter-city bus services do now operate in that range. The average length of haul is, however, only one-fiftieth that of the railways, but this is offset by the much greater volume of originating tonnage for road haulage. Average passenger journeys are also proportionately shorter than on the railways, with a mere 2–3km in urban passenger transport. Undoubtedly the powerful railway lobby retarded development, but it is now accepted that road transport will grow as vehicle manufacture expands. It is nevertheless criticised as demanding higher labour inputs than other forms, important where labour shortage exists in many districts, while Soviet planners have opposed any policy that might lead to an explosion of the private motoring sector, regarded (perhaps rightly) as a monstrous waste of national resources.

Three-quarters of total road freight traffic and two-thirds of road passenger traffic occurs in European Russia, where the large towns are particularly significant foci of road traffic. One-third of all road traffic takes place in the central industrial and central black earth regions, in the industrial south and in the Ural. Road transport plays a notably important role in oil-producing districts and in the northern forest lands. Although road transport is only of the most modest dimensions in the eastern regions, it plays a vital role away from the railheads and river wharves. In some instances roads of this type have ultimately been converted into railways when traffic volume has reached critical thresholds. Urban bus services and those running into the adjacent countryside from towns occur throughout the national area, while many towns have efficient and expanding tram systems. Inter-city bus links appear most significant in southern Siberia and in Central Asia, though there are considerable networks on the good road systems of the Baltic republics, the Ukraine and in Caucasia.

Compared with Western Europe, the comparative length of Soviet 'motor roads' at three times the length of the railways is low: in the United Kingdom, the comparable figure is nineteenfold! Of 1.4 million km of 'motor road' in 1978, some 741,600km were hard-surfaced (a 60 per cent increase in ten years). Grit or dirt roads can turn to quagmires in early winter and in the spring thaw; in winter they are rough and rutted but frozen, and in summer they form ribbons of dust. Even the metalled roads suffer acutely from frost-heaving and are particularly difficult to maintain wherever there is permafrost or its likely occurrence. In the northern forests roads are commonly of the *corduroy* type, a log structure covered by grit and suited to heavy vehicles. Nevertheless, the hard frozen surface of winter encourages road use and in Siberia 'winter roads' appear across otherwise impassably wet terrain, while even frozen rivers are used as roads. Intense winter cold presents, of course, problems for the internal-combustion engine through increased viscosity of fuel, freezing of cooling water and hardening of rubber tyres, as well as excessive heat loss from engines.

It is, however, possible to talk of a true road network only within the well-settled triangle of European Russia and Western Siberia, though local networks are found in the more thickly settled parts of Caucasia and Central Asia. Simple local dendritic forms mark most of Siberia and the poorer parts of Central Asia. No major motor road yet crosses the country from west to east and information suggests that it is still impossible to motor right across Siberia to the Pacific coast. In the Amur basin disconnected road systems focus on the railway and run several hundred kilometres northwards into the tayga, as, for example, the Aldan Highway that serves Yakutsk. In northern European Russia and in northern Siberia local road systems focus on railway stations or river quays, but there are some lengthy and probably very rough roads acting either as portages between rivers or joining together remote settlements, with one of the most extensive systems of this type serving the mining settlements of the Kolyma basin from the port of Magadan.

8.5 Pipelines

The first pipeline across the Caucasian isthmus was laid before the First World War, but the real expansion came in the 1960s. This has included building long-distance lines to carry oil and gas to Eastern Europe, but also some long internal lines to serve Siberia and a notable system of gas lines focusing on the central industrial region around Moscow is beginning to emerge. Pipeline transport accounts nowadays for about a fifth of all transport effort. Although relief poses no special problems to pipeline construction, there are difficulties in laying lines across extensive swamps or across some of the excessively gullied areas of the south, while operationally extremely low winter temperatures demand effective insulation, and pipelines across permafrost are liable to distortion and rupture through upsetting the local thermal balance.

8.6 Air transport

Since the 1920s the Soviet Union has been remarkably air-minded and all civilian air transport is run through *Aeroflot*, the world's largest operator, with well over 500,000km of trunk and inter-city routes. The speed of air travel on the major axial routes makes substantial savings on travel time and cost compared with rail travel and, as an example, as much as a man-week can be saved through using the Moscow–Vladivostok flight. In the desert and arctic regions aircraft have become the prime-mover for passengers and make a major contribution to freight transport, with over two-thirds of all air freight moving in the roadless north and east. It is claimed that equipping an air route (including airfields, navigational aids, etc.) is only one-tenth as expensive as a well-built road or railway in these remote areas of great physical difficulty. The challenge for air transport in this type of setting undoubtedly encouraged the Soviet aerospace industry to develop heavy-duty helicopters, such as those capable of lifting fully laden vehicles, while there has been talk of using airships. It should be remembered that Soviet flyers have great experience in Arctic flying, with one of the first flights across the North Pole made in 1938. Air transport seems an ideal medium in the long periods of still anticyclonic conditions over much of Siberia, when ground conditions for land transport are tedious.

The major obstacle to developing air transport in the inter-war period, despite an actively encouraged air-mindedness, was the limitations of the Soviet aircraft design and building industry, though machines were bought in from outside (particularly Germany) in the early 1930s. The rapid rise of air transport may be correlated closely in time with wartime needs and post-war experience using German technology captured in the closing stages of the war or immediately thereafter, while American expertise had been received through wartime aid. The fortuitous human element also presented itself at this time in designers such as Antonov, Tupolev and Ilyushin. Certainly, Soviet territory, lying astride great circle routes from Europe to Asia, could become a focal point of world air traffic if the Soviet authorities were to allow more foreign operators across their airspace.

8.7 Postal- and telecommunications

Until the telegraph became a vital means of government in the mid-nineteenth century, the tsar's couriers were often weeks or even months in making their way to remote provincial towns. In the 1860s there were even moves made to link Europe to America via Siberia, the Bering Strait and Alaska, but the plans were overtaken by the laying of the Atlantic cable. Since the Revolution the telephone network has been widely extended, and nowadays almost all village soviets, state and collective farms have telephone communication, but the scanty statistics suggest that telephones are disproportionately located in towns (84.7 per cent). Radio telephones are also used in the more remote areas, including links with bases on the Arctic islands. In its postal system Russia lagged behind Europe in general, with the first adhesive postage stamps issued only in 1858, but the institution of airmail in 1923 was a major advance and a considerable proportion of the mail now moves in this way.

The Soviet period has also seen considerable attention given to all means of regulating and influencing public opinion. Radio transmissions cover the whole country, and since the mid-1960s there has been expansion of the television service over the main populated areas and to centres in northern Siberia. Originally relay stations were built to spread the coverage of reception, but nowadays satellites are employed in telecommunications. Use of systems such as 'Prestel' for information dissemination could be of immense importance, just as electronic data-gathering and processing using widely scattered terminals could have great administrative impact. Communication is, however, bedevilled by the wide time difference across the country, with virtually a working day's time difference between European Russia and the Far East, with a two to five hours' difference between Moscow and the principal towns of the Ural and Siberia.

The traveller between the principal towns of the Soviet Union will be quickly aware that his journey is being made along broad lateral corridors of communication. If he travels by train, he will see paralleling the railway, sometimes close, sometimes further away, a major highway; he will notice lines of telegraph poles and electricity pylons stalking the route; he may even glimpse at one or two points a pipeline and now and then a television relay tower. Overhead, he will notice aircraft obviously routed along some similar corridor. Study of reasonably detailed Soviet maps of these 'means of communication' will confirm this pattern, nowhere more strikingly than in Siberia, the Far East and Central Asia, but also even to a lesser degree in northern European Russia, while a clear hierarchy emerges over the main triangular settlement area (*ecumene*). The spatial and strategic implications of such a pattern need no elaboration and reflect interestingly upon the nature of Soviet space.

8.8 Where to follow up this chapter

The most useful survey of transport is Symons, L. and White, C. (eds), *Russian Transport: An Historical and Geographical Survey* (Bell, London, 1975), while Hunter, H., *Soviet Transportation Policy* (Harvard University Press, Cambridge, Mass., 1957), and Hunter, H., *Soviet Transport Experience* (Brookings Institution, Washington, D.C., 1968), are two still valuable sources.

The classical Soviet work is Nikolskiy, I. V., *Geografiya Transporta SSSR* (Moscow, 1960, 1972, 1980), but also important is Kazanskiy, N. N. (ed.), *Geografiya Putey Soobshcheniya* (Moscow, 1969), and Galitskiy, M. I. *et al.*, *Ekonomicheskaya Geografiya Transporta SSSR* (Moscow, 1965).

9

The Soviet Union and the World since 1945

The years between the world wars saw a deep mutual fear prevail between the USSR and the world at large, where the former had few, if any, real friends. In the Soviet Union the rising xenophobia of Stalin and the Politburo engendered isolation and secrecy, with subversive plots suspected in every action that diverged even marginally from the prescribed party line. Fearful of international Communism, the countries of the outside world threw a tight *cordon sanitaire* around the innovations in society and the economy being applied within the USSR. Neither side knew or even wanted to know much about the other. The outside world grossly misjudged, both by over- and under-estimation, the real achievements of the new Soviet regime, while the Soviet political bosses, with almost no experience beyond their own boundaries, gravely misjudged the true state of capitalism and, in particular, how to conduct relations successfully with the world at large. From both sides reluctant dealings were dominated as much by ineptitude and clumsiness as by mistrust and suspicion.

Consequently the most unexpected aspect of the Second World War was perhaps the success of the Soviet resistance to the German invasion and certainly the least anticipated outcome in 1939 was that the end of this global conflict would see the Soviet Union on a political and military par with the most powerful of the victorious powers.

Through the advance of the Red Army most of Eastern Europe was quickly absorbed into a new Soviet *imperium*. Shortly after the German collapse, as Japan was crumpling under the American atomic onslaught, the Red Army invaded Manchuria and took charge of this vital gateway to Siberia and the warm-water ports of the Liaotung Peninsula. Moscow could now act as tutor to the embryonic Chinese Communist state. The political and military strength of this position was not matched by economic and technological strength within the Soviet Union. The immediate post-war years had therefore to be a harsh time of repairing the destruction left across the territories that had fallen into German hands and at the same time building a more powerful industrial base adjusted to the locational shifts in industry generated by the wartime 'scorched-earth' policy that had carried so many plants deep into the interior from European Russia. Most critical of all was the need to close the technology gap between itself and the West, especially as the USA at the time held a decisive lead in nuclear weaponry and air power. Not only was Soviet science and technology involved, but the scientific intelligentsia of the new Eastern European satellites were also harnessed to the task, while mistrust and suspicion between the wartime allies were generated by a ruthless Soviet effort to obtain technological and scientific knowledge through espionage. The Soviet Union again returned to its secretive and closed economy and society, withdrawing from the world as the xenophobia of the ageing Stalin intensified and closed over the new satellite *imperium* in Eastern Europe, where the individual states' economies were ruthlessly tied to the Soviet Union's needs.

The rehabilitation and extension of the Soviet economy was slowed, however, by the maintenance of massive armed forces that sucked up scarce manpower from that remaining after the appalling war losses, for in the aura of fear and mistrust now pervading both the USSR and the West neither felt it could safely demobilise beyond a limited degree, so the post-war years became a period of 'cold war', a nervous armed truce. Trading on the Western fear of these large Soviet forces, the Americans concentrated their efforts on further isolating the Socialist *bloc* and on strengthening their position in a ring of containment around what appeared to be a still expansive Soviet Union. Some students suggest the USA did not wish to reduce tensions too quickly from war to peace, because the economic reorientation required could have had disastrous domestic consequences. On the other hand, the Soviet Union certainly sought to create diversions through 'small wars' and provocative action, hoping through one or other to win a massive prestige victory. So long as war could be restricted to conventional weapons without the use of strategic air power, the Soviet position was a remarkably strong one.

During the 1950s, once Stalin had died, the Soviet leadership modestly changed direction and emphasis. Certainly, through a quite remarkable effort, the Soviet Union has managed to close the military-technology gap between itself and the USA, even, in fact, to take an initial lead in space technology. In one way this greater equalisation made the rimland between the Soviet and American spheres of influence less a zone of confrontation, but with nuclear weapons and long-range missiles in the hands of both super-powers, the one-time strategically negative Arctic is now crossed by the vital great circle trajectories across which missiles may potentially be lobbed at each other by the super-powers. Nevertheless, during this same period growing internal pressures forced Soviet leaders to think seriously about domestic questions, because as the rehabilitation of the economy proceeded and became increasingly apparent, the population began to expect a betterment of living standards. It was consequently necessary to dismantle the Stalinist siege economy, and this also meant reducing tension with the world outside. By the early 1960s the sophistication of the economy had reached a threshold where acceleration of further growth and the necessary technological advance could best be achieved by buying in know-how from others, including from the West. Demographic patterns were creating a growing shortage of labour, so there was further reason to reduce international tension to allow manpower to be released from the armed services. The Soviet Union was also becoming concerned with the trend of events in China, whose relations with its former tutor were quickly deteriorating as it became economically and technologically stronger.

The 1960s saw Comecon become an effective economic organisation within the Socialist *bloc*, though only at the price of concessions to the Eastern European satellites for more room to manage their own affairs, achieved by insurgency in some countries, that undoubtedly damaged the Soviet image, while in others relaxation of the Soviet grip was managed 'behind closed doors'. The Soviet invasion of Czechoslovakia against an excessively far-reaching economic liberalisation there in 1968 brought a powerfully adverse reaction, overt or veiled, among both Comecon members and the outside world that certainly called for added caution in the Soviet Union's relations with its neighbours. Comecon (the Council for Mutual Economic Assistance) had been formed in 1949 as a showpiece counterweight to the American Marshall Plan aid and to moves towards international collaboration in Western Europe that culminated in the European Economic Community. In Stalin's day Comecon had been little more than a name, but in the ensuing changes its members outside the Soviet Union, though stoutly Marxist-Leninist, had increasingly pressed their own point of view, often far more sophisticated than Soviet experience. Forced in

Stalin's day into a straightjacket of imitating Soviet practices and ideas, they have also sought to apply Marxist-Leninist dogmata more to their own special conditions and have succeeded in making changes towards more elaborate and advanced practices in trade, banking and monetary mechanisms, all relatively primitive and doctrinaire under Stalin. These comparatively small and mostly far from wealthy countries could least afford to pursue the self-sufficiency policies so beloved of Stalinist planners and they have done much to encourage intra-bloc trade, weaning the Soviet Union away from its former self-contained and introvert ways.

9.1 The Soviet trading pattern

In Stalinist economics, with every emphasis on a self-sufficient closed economy, trade was of little interest, carried on only for essentials not available within the USSR. On offer to the outside world were raw materials such as ores and petroleum and agricultural products like coarse grains. Soviet business practice conducted trade as a simple barter operation, but at the low level of transactions carried on this was not necessarily a serious drawback. Trade was, however, already seen to have a political significance, with this consideration often influencing the terms on which deals were made. The Eastern European nations drawn into the Soviet orbit after the Second World War were, in contrast, used to being actively involved in the financial and business patterns of Europe's multilateral trade. In the early post-war years several Eastern European countries had well over 60 per cent of their trade with the Soviet Union and all were forced to give it preference under conditions invariably biased in the USSR's favour. This system the Soviet authorities regarded as simply an extension of their own closed system and would have liked to have forced their complete dependence on the Soviet Union because of the political-geographical advantage of making them as subservient as possible in a system designed essentially to serve the USSR's own economic rehabilitation. Although pressurised by the Soviet Union to pursue similar economic strategies to its own, the idea of self-sufficiency and self-containment was completely unreal in relation to their individual resource endowment and the sizes of their own domestic markets.

In the main the Soviet Union sought manufactured goods from Eastern Europe, favouring trade with the more industrialised countries like the German Democratic Republic, Poland, Czechoslovakia and even Hungary. In return, raw materials and fuel were sent to enable more goods to be made for the Soviet Union with a little to spare to allow the satellite countries their own slow rehabilitation.

Specification and design was increasingly influenced by Soviet needs, and sometimes, as in the German Democratic Republic, whole plants worked solely on Soviet contracts. The less industrialised Eastern European countries came to occupy a rather secondary place (with discontent at this role most vociferously expressed by Rumania), unless they produced raw materials or agricultural produce of special advantage to the Soviet economy. Some of the goods received from Eastern Europe were re-exported under Soviet labels. Eventually, however, even the Soviet authorities admitted they had been paying unrealistically low prices for such goods.

The relaxation of the closed economy after the death of Stalin was reflected in growing pressure from the Eastern European members of Comecon for Soviet commercial and financial practices to come closer into line with those outside the Socialist *bloc*. At the same time, trade within Comecon and between it and the West was diversified as the Eastern European countries abandoned futile efforts at a self-sufficiency impossible for them to achieve. The move to specialisation in Comecon, in which each country produced what it could do best and then traded the surplus of these articles with other members for their special products to make up its own deficiencies, was in itself an encouragement to trade. Although the proportion of intra-Comecon trade of the individual members has declined, trade between the Eastern European countries and the USSR still amounts to around half the former's total. It is, however, now clear that the Soviet Union in certain critical sectors would like to reduce its commitment to its Eastern European Socialist *bloc* neighbours. For example, they have been told to look for alternative sources of energy imports, since the Soviet Union does not intend in future to continue to regard them as 'favoured nations' for supplies of petroleum and natural gas.

Increasingly faced by hurdles in making economic progress, the Soviet attitude to trade changed during the 1960s, especially in seeking contacts with Western industrial nations. To cut down the lead times in development, the Soviet state trading organisations began to look to the purchase of specialised and sophisticated equipment and know-how from the West. The Soviet Union has offered raw materials such as ores and coarse grain or timber as well as petroleum and natural gas, but there has also been a growing trade in some manufactured goods from its factories. The Soviet Union has also bought consumer durables, so that it can save on its own investment in the necessary production facilities and yet provide an incentive to Soviet workers to work harder to earn the money to buy such desirable items as refrigerators and washing-machines. Purchases have also included large quantities of clothing (including

Table 9.1 *Soviet foreign trade (in million roubles at contemporary prices)*

	Export		Import	
Country	1970	1978	1970	1978
Bulgaria	844	3,145	973	2,997
Hungary	758	2,396	722	2,430
German Democratic Republic	1,738	3,982	1,557	3,711
Cuba	580	1,947	465	2,222
Mongolia	178	596	53	147
Poland	1,215	3,450	1,135	3,600
Rumania	445	971	474	979
Czechoslovakia	1,083	3,002	1,110	3,059
Yugoslavia	294	1,108	226	1,070
Austria	67	371	88	272
Belgium	74	408	75	257
United Kingdom	418	857	223	668
Italy	191	1,112	281	859
Canada	7	29	118	359
Netherlands	151	300	72	162
USA	58	253	103	1,599
German Federal Republic	223	1,363	321	1,941
Finland	258	1,004	273	1,186
France	126	840	287	974
Sweden	105	312	130	183
Japan	341	736	311	1,584
Algeria	62	88	56	52
Afghanistan	36	139	31	76
Brazil	2	35	21	130
Egypt	327	148	279	198
India	122	364	243	407
Iraq	60	674	4	410
Iran	169	433	62	238
Turkey	56	89	27	69

Source: *Narodnoye Khozyaystvo SSSR* (1978 edition).

women's fashion wear) and shoes. The quantities of such consumer goods bought each year can be varied to act as a regulator in the Soviet domestic retail market. Although some of the goods have come from Western Europe, most have been supplied by the more industrialised Comecon members like Czechoslovakia, the German Democratic Republic, Poland and Hungary.

Although a state monopoly on trade through the central purchasing agencies has been retained, the Soviet Union has moved closer to the financial and commercial practices of the West and has consequently increased trading possibilities. Nevertheless, no effective multilateral system with the Socialist *bloc* has yet been developed. It is perhaps not without significance that the German Democratic

Republic is the Soviet Union's main trading partner in Comecon and the German Federal Republic has the same role among the Western nations. Both German states are, of course, highly industrialised countries, but there is the important fact that the Germans have long experience of the Russian market dating back to tsarist times and were one of the few traders with the early Soviet state between the wars. Trade with North America has been constrained by the American reluctance to sell any goods of a potentially strategic nature to the Soviet Union. Nevertheless, in some years the USA and Canada, along with Australia, become important suppliers of bread grains, made necessary by failure of the Soviet grain crop. The European Economic Community has also sold some of its agricultural surpluses (notably from the 'butter mountain') to the USSR, though the deals have been severely criticised within the EEC because of the unwarrantably low prices charged.

In dealings with the developing countries the Soviet Union has faced the problem of having to accept commodities it does not always want. On occasions these have been re-exported from the Soviet Union, though this practice has been vigorously attacked by several Third World countries. A particular example was that of cane-sugar imports from Cuba that forced a reduction in the area sown to sugar beet in the Ukraine but also resulted in a surplus of sugar in the Soviet Union that was difficult to dispose of. One of the greatest opportunities for development of trade lies with Japan, desperately in need of many raw materials found in Siberia and able to supply high-technology goods in return.

Year-to-year trade patterns are also affected by political interests. As noted, trade has been limited by the USA's reluctance to sell a wide range of goods to Socialist *bloc* countries because the items may have strategic potential. The USA tried, however, with only modest success to get other members of NATO to join it in this action. Western observers have suggested that some Soviet manufactured goods like cameras, watches, radios and even motor-vehicles have been sold in Western markets at unrealistically low prices, while Soviet sales of petroleum at unusually modest prices to some smaller Western European countries have been seen as attempts to get a stranglehold in small but important markets.

Since the late 1960s the Soviet Union has bought complete installations directly from Western manufacturers, like whole factories or integrated systems of computers, clearly designed to help accelerate economic development. Attempts to get the relatively poor Eastern European countries to invest in joint development projects through Comecon were not outstandingly successful, so the idea has consequently been hawked around in capitalist countries, though proposals for deep financial commitment have been received with caution. The projects involving Eastern European members of Comecon have included the sharing of the construction of pipelines to carry oil and gas from Soviet fields, but there have also been industrial projects like the Ust-Ilimsk and Usogorsk cellulose and paper combines, the Kiyembayevo asbestos plant and works associated with the *Kursk Magnetic Anomaly* ores. Poland and the USSR also have a joint engineering plant at Novovolynsk, and the Soviet Union has had Eastern European help in planning and developing nuclear power stations. Western countries, although they have offered the Soviet Union long-term credits, have been careful in taking up projects. Japan, seriously concerned about its raw-material supplies, has with some trepidation interested itself in joint agreements over the working of oil in Sakhalin, coal in southern Yakutia and the large Udokan copper deposits. In 1973–4 West Germany explored various options on energy supplies from the USSR, while even the USA showed a brief interest in a similar deal. Some long-term credits have been taken up by the Soviet authorities to have international consortia build whole plants for them: for example, a plastic-making plant in Northern Caucasia, an ethylene plant at Omsk, and part of the steelworks at Staryy Oskol near Kursk, among others. One major deal was the building of the Tolyatti motor works by Fiat of Italy, while the vast Naberezhnyye Chelny lorry plant has had an international involvement.

The first few of these deals were on a barter basis through payments to the consortia in goods, so that the Western companies participating had to find firms able to dispose of those items which they themselves were unable to market. There arose the so-called 'switching' deals, but this was a cumbersome system and encouraged Western consortia to offer long-term credits guaranteed by their governments. The more recent stage has been to sign 'compensation' agreements, where Western consortia provide the plant and know-how in return for a share of its production for a specific period, usually ten to fifteen years. For example, Austria, West Germany, France and Italy have supplied large-diameter pipe used to build pipelines to supply natural gas from Soviet fields to Western Europe. Some Western sources are concerned that the compensation system will bring unwanted manufactured goods into the West that could upset delicately balanced market forces. It is also argued that by selling large-scale, highly productive plants, the West could kill lucrative markets for their own products in Eastern Europe.

Certainly, it could be argued that the pace of development of technology in the West has been such that the Soviet Union could not have hoped to have kept abreast without purchasing Western know-how and equipment. The immense research and development infrastructure of the huge Western multinational firms has no true equivalent in the Socialist *bloc*. The exchange of ideas and patents within Comecon on the basis of 'no secrets among friends' by state-directed research does not seem to give comparably powerful results, particularly in such key areas as electronics, computer technology and micro-processors, but also even in industries like petrochemicals. Probably too much of the Comecon research effort and its development facilities were for too long committed to overtaking the USA in the prestigious and strategically significant but otherwise limited space and missile programme.

A Soviet dilemma is where to direct its trading effort, for political and economic objectives clash in many instances. The greatest difficulty arises over trade and development aid for Third World countries. A few are members of the Socialist *bloc*, but the majority are regarded as part of the capitalist system, which accordingly colours relations with them. In some instances Soviet approaches have been rebuffed by Third World countries when they have felt unduly pressurised to imitate and accept the Soviet interpretations of socialism. The Soviet Union has been vociferous in supporting demands for development aid for Third World countries from the Western industrial nations without feeling any real obligation itself and without itself setting any particular notable example. In its contacts with the Third World the Soviet Union has stressed the need for them to be of 'mutual advantage', which in effect means that aid without strings is impossible.

Soviet aid in credits and gifts has been modest, probably amounting to only about one-twentieth of those made by Western countries. The commitment of the GNP to the Comecon countries in total has only been a quarter of the level of commitment by the West. Scientific and technical aid has also been mediocre: a few major projects like Bhilai, Heluan, Aswan and the Iraqi dam are cited with almost monotonous regularity. While much of the aid given to Third World states concentrated on a few major projects of potentially great propaganda value, any faulty planning and other shortcomings in the undertakings have quickly received adverse publicity and criticism, backfiring on the Soviet image. There has, however, been a considerable educational contribution, possibly because of its important political undertones and because it can be given at home in the donor's territory. Aid has in general been limited to relatively few countries, notably those whose governments appeared to be favourably inclined to follow a 'socialist path' or where influence was desirable for strategic reasons, while several countries, particularly in Africa, were included to counter Chinese efforts to win their

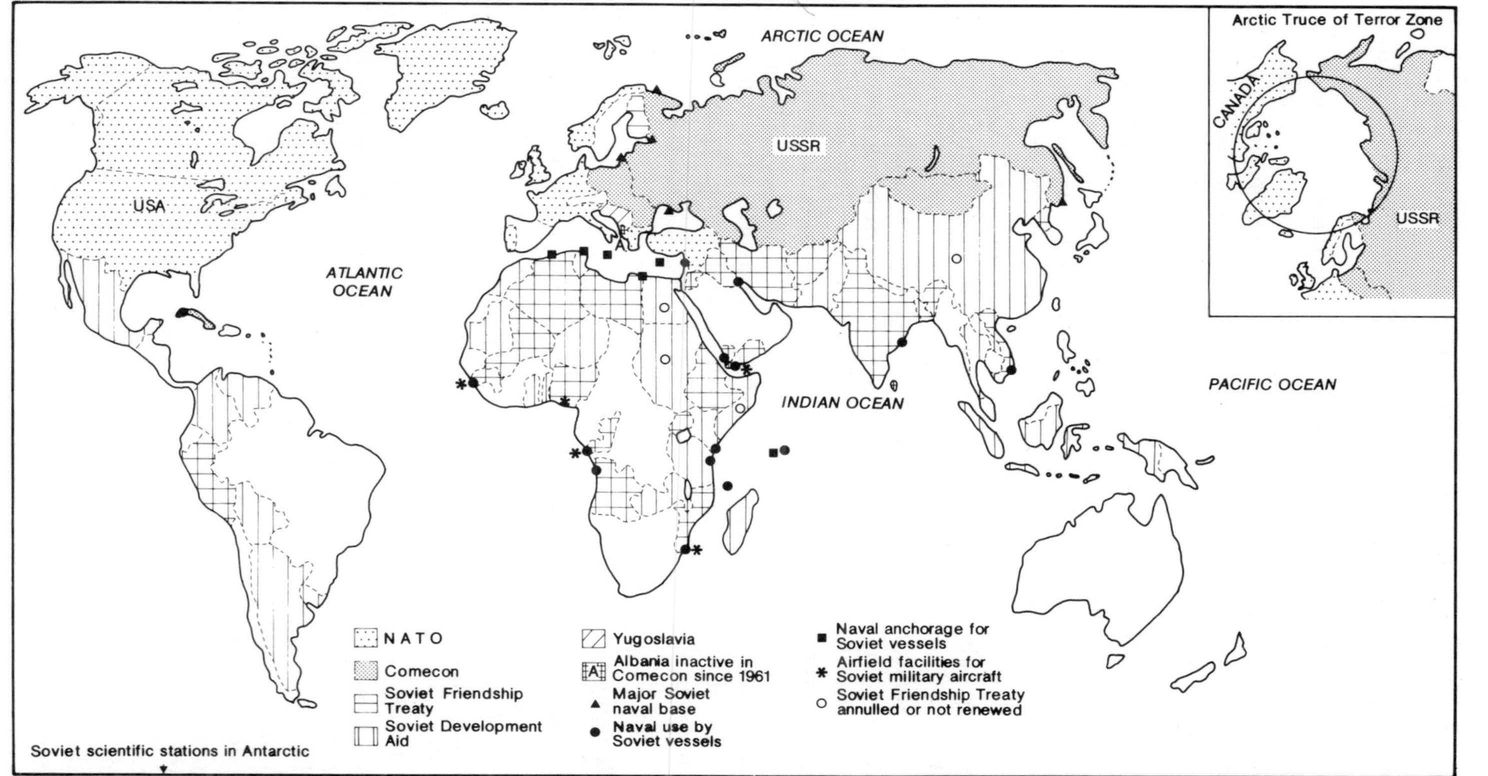

Soviet political and economic links are now world-wide and there is a formidable array of naval facilities for Soviet ships in African and Indian Ocean ports.

Source: updated from a map in *Informationen zur Politischen Bildung 182* (Bonn, 1979).

Figure 9.1 *The world distribution of Soviet aid*

support. Not all the recipient countries have remained faithful friends of the Soviet Union.

Soviet Third World aid and trade have suffered through the political instability of many of these countries, just as this instability has also been a factor in encouraging the industrialised nations of the Western world to moderate their own aid and trade with the developing countries. Although not without its dangers, for the industrial nations of the West the higher higher level of stability in Comecon markets has been an attraction, especially as the goods the West has to offer find a more receptive market in Comecon than in many Third World countries. Trade has been encouraged, because Western businessmen have been impressed by the prompt payment and honouring of agreements by the Soviet Union and other Comecon countries, so long as the contract is adhered to carefully. Soviet and Comecon goods, designed and produced in a simple and robust fashion for their own markets, find a more ready acceptance in the Third World than refined and sophisticated Western products often unable to stand up to the rough conditions of most less developed countries. Nevertheless, the Soviet Union finds trade with the Western industrial countries much more attractive not only because they have much more than the Third World to offer that it needs but also because trade with the industrial West brings in urgently needed hard currency.

9.2 Changing spatial patterns in the USSR

As its involvement with Comecon has grown, there has been a change in spatial emphasis within the Soviet Union. A new interest in the economic development of westernmost European Russia, notably Byelorussia, has become apparent. Once a marchland against an unfriendly outside world, these districts are now conveniently located to develop closer associations – especially industrial ties – with the economically more advanced members of Comecon, like Poland, Czechoslovakia and the German Democratic Republic, while they are also well positioned to trade with the outside world through Baltic and Black Sea ports. As the Soviet economy has become less introvert, the advantages of concentrating development in the deep interior has also proved to be less attractive, despite a continuing heavy commitment to the eastern regions in the tenth Five-Year Plan. There is evidence to suggest that inter-war Soviet economic growth was retarded by the massive inputs needed for capital-demanding projects in Siberia where the harshness of the environment added a formidable development premium, and yet such projects were unavoidable when such an inward-looking approach to a closed and self-sufficient economy and society was made. There were at that stage, however, several strategic attractions as the interior lines of communication and the interior location of so many major projects were well beyond attack until the age of long-range airpower. Moreover, the deep interior offered much to an economy that was in the basic stages of simple capital construction of a national infrastructure, for the necessary raw materials were conveniently near to where the new industrial structure was being built. As technology has shifted towards more labour-intensive products with high research and development demands, the more populous parts of the Soviet Union, already possessing a sophisticated infrastructure, skilled labour pools and close spatial possibilities of industrial interlinkage, have grown in attraction.

One most obvious role for Siberia is as a storehouse of raw materials, especially energy, but with increasingly difficult deposits to work and rising capital costs for development of energy resources, there is a growing Soviet interest in attracting outside help to contribute both technology and investment. The Eastern European Comecon members have little to spare for such action, but already Japan has made a contribution and other Western nations have shown interest. The opening up of raw-material sources will depend on world price levels, for if these rise significantly Siberian wealth becomes more attractive. With the energy problem probably the greatest challenge to the world in the coming decades, hopes that the Soviet Union would become a major supplier of petroleum to world markets, notably from Siberia, now seem ill-founded and it is even possible that the Soviet Union could become instead a net importer in the early 1980s. Despite the large discoveries in the west Siberian lowland, the long-term needs of the Soviet Union have been growing substantially and also likewise those of its Comecon associates, whose supplies have so far come overwhelmingly from the Soviet Union on a 'favoured-nation' basis. The situation for natural gas seems more hopeful, though many of the main deposits lie outside Siberia. It is perhaps Siberia's immense potential for hydro-electric development that makes the most exciting possibility for the future, either through transmission of current to the workshops of Europe or the siting in Siberia of energy-intensive industries. Siberia may also be acceptably remote enough for the building of giant nuclear power stations; indeed, since 1973 Comecon's *Interatomenergo* has been fostering the use of nuclear power and the Soviet Union is well advanced in fusion technology.

Siberia may also play a role in the relations between the two ocean-related commercial-industrial spheres of the Atlantic and the Pacific. The latter has been particularly distinguished by the massive new grouping of commercial and industrial activity in the western USA and in Japan, while most recently powerful trading foci have emerged elsewhere in Eastern Asia, like Taiwan, South Korea, Singapore and Hong Kong. The Atlantic sphere is composed of longer-established industrial districts, notably the American Atlantic Coast and Mid-West (whose accessibility has been improved by completion of the St Lawrence Seaway) and the industrial concentrations of Western and Central Europe, now augmented by the emergent industrial nodes of Eastern Europe. Although trade between the Pacific West and the Atlantic East – between Japan and its neighbours and Europe – can and does pass across the American meridians (notably seaborne through the Panama Canal), a more intensely used route is that of the Asian periphery, particularly when the Suez Canal is open. Such sea routes are, however, slow, and far more attractive is the relatively short haul by the virtual great circle route across Siberia. As trade shifts from low-grade bulk goods to light, small-volume, high-value manufactured goods, the attractiveness of the land route grows, made even more attractive by the development of containerisation, allowing easy transhipment between transport media and appreciably reducing terminal costs. Already the Trans-Siberian Railway handles a rising volume of containers, and completion of the Baykal–Amur trunkline as a full alternative will make this axis of shipment even more worth while. Leading into the backyards of European industry, it is considerably shorter and simpler than the corresponding sea–land–sea route via North America. Furthermore, crossing Siberia as it does, it makes possible a range of intermediate sites at likely price-sheds in European–Asian trade, presenting great opportunities for international investment, particularly for capital-intensive rather than labour-intensive industry, since labour is one of Siberia's scarcest commodities.

The effective peopling of Siberia is not just a question of providing labour, for there are strong political-geographical undertones. As China's population grows and that of Siberia remains almost static, the Soviet Union must feel a threat to its own position, particularly as China has expressed interest in Soviet territory. Chinese settlement on a massive scale early this century in Manchuria was conspicuously successful and no doubt could be expanded into what the Chinese would regard as a grossly underpopulated Siberia. Consequently, the Soviet Union, while encouraging growth in the more sophisticated parts of its economy in its western territories, would equally welcome any development that gave a stimulus to settlement in Siberia through the incentive of making life there more attractive. Such could be achieved through diversified industrial development in the towns of the railway zone. Although in terms of the Soviet economy investment in Siberia has become less effective in unit terms than in European Russia (where there are fewer infrastructural problems), this might be rectified by putting Siberia more into the world setting. For a multinational development, even the high costs of Siberian development could well prove attractive in locational terms on the Pacific–Atlantic continental price-shed. If this were to strengthen the impression of Siberia as a Soviet land, it would prove an almost irresistible bait for Moscow. A study group of the North Atlantic Treaty Organisation has seen Siberia as being potentially capable of becoming richer than Canada. While this must have pleased Soviet leaders, it may also remind them that Canada did not reach its present affluence within a closed system of the British Empire, but rather because it was open to international investment and settlement.

9.3 The Socialist *bloc* in the world

The easier relations between the Socialist *bloc* and the outside world compared with the years of the 'cold war' have in themselves also brought changes in the spatial pattern of the economy of the Soviet Union as trade has opened up. Although tensions have not disappeared, some of the greatest change has come in relations across the Iron Curtain, which has in effect become an expanded metal grille, the size of the mesh opening and closing as the international climate warms and cools. Unquestionably much of the drift towards a more normal situation has arisen from improved relations between the individual countries of Western and Eastern Europe and has gone forwards as the Eastern European members have strengthened their own positions within Comecon. That the Soviet Union has tolerated such improved relations may be seen perhaps as a necessary means of holding these Comecon members within the Soviet sphere by at least allowing them some semblance of management of their own affairs in foreign relations, though this was sorely tested in Poland in 1981. Particularly significant has been the change in the attitude of the Soviet Union towards the European Economic Community, and the initial uncompromising hostility has mellowed to a position of preliminary moves towards talks with Comecon on matters of joint interest. One obvious reason is, of course, that for the Eastern European countries Western Europe has always been one of their most promising markets, while they have equally traditionally drawn many of their needs from this trade area. For the Soviet Union prospects of better relations with Western Europe are attractive in making easier receipt of technical aid or even investment in selected projects.

Perhaps, however, the search for acceptable ways of dismantling tensions along the European flank has been conditioned by what appears a far more difficult situation on the Soviet Union's eastern flank, where long-standing if currently latent historical frictions over territory exist between the Soviet Union and both China and Japan. Soviet leaders doubtless sense that efforts to involve Japan through investment and technology in Siberian raw materials, so vital for both the Japanese and Soviet economies, may be an effective 'neutraliser' and that such involvement may deter any Sino–Japanese aspirations to a *rapprochement* damaging to the Soviet Union. China holds a particularly strong strategic position in the Asian interior and its present efforts to colonise the empty western territories of Sinkiang and strengthen its position in Tibet must disturb the Soviet General Staff. From the Soviet viewpoint, in the long term a *rapprochement* with China could bring the greatest benefits and would by its nature almost inevitably draw in Japan, whereas from the Western point of view such a *rapprochement* could be regarded as strategically and economically most unwelcome.

9.4 The Soviet dilemma in the Third World

The difficulties faced by the Soviet Union in giving aid to and trading with the less-developed countries have already been outlined, but it is perhaps worth considering the broader political-geographical implications of Soviet relations with the Third World, particularly as the Soviet authorities see trade and aid as only worth while if done for 'mutual advantage'. Territories seeking freedom from imperial ties or from otherwise oppressive regimes offer obviously great opportunities to introduce a 'socialist path'. To be a credible champion of such causes, it must be possible to establish as direct a contact as possible. One of the claims the Soviet Union might make to emergent countries is that its own experience is particularly applicable to the problems faced by these countries.

One key political-geographical reason for Soviet interest is that a considerable number of Third World countries lie in the broad buffer zone around the USSR commonly termed the 'inner crescent' or the 'rimland' by students of Mackinder's thesis on the geographical pivot of history. Soviet interest has also extended to newly emergent countries with socialist tendencies in North Africa, because if events were to follow certain courses these might be useful for penetration of southern Europe. A key position was identified in Egypt, lying across the landbridge between Africa and Asia and controlling the

Suez Canal. Unfortunately, Soviet aspirations here failed, though important political influence in the Arab world can still be exercised through Syria and Iraq. South Yemen is another important base for Soviet interest in the Horn of Africa, where Ethiopia has been drawn into association with it, and which for a time also extended to Somalia. Relations with Islamic states must be seen as part of a Soviet need to preserve its own internal stability, with considerable Islamic populations in Central Asia and reaching as far north as Kazan on the Volga, probably numbering at least 35 million. The issues have become more complex as a new militant Islam appears to be rejecting 'westernism', directed at both American and Soviet forms. Soviet overtures, despite their need for aid, were rebuffed by both Turkey and Iran, two countries with considerable mistrust of Soviet aims. The fact that Soviet troops long overstayed their welcome after the Second World War around Tabriz was not overlooked by the Iranians. Far more persuasive have been Soviet overtures in Afghanistan, leading to a Soviet 'protectorate' being established early in 1980.

In the 1920s the Soviet Union managed to establish a Communist People's Republic in Outer Mongolia, one of the most backward parts of Asia, that has been used as a showpiece of what Soviet help can achieve. Between the world wars considerable Soviet help was given to Communist insurgents in China, notably in the south-west, but Soviet influence in Sinkiang was also strengthened. After 1945, with the Soviet authorities holding the former Japanese base in Manchuria, massive aid was given to the Communists, and no doubt was a major contribution to their eventual triumph. Through the 1950s China welcomed Soviet aid as a means of rehabilitating its ruined economy, but political relations deteriorated as the difference in ideological interpretation between the two countries grew. The Chinese felt the Soviet Union had left them with intolerable burdens in the Korean War and had failed to support them adequately in the dispute over the Indo–Chinese border and in solving the 'Taiwan problem'. From the late 1960s Soviet–Chinese contacts ceased and the former friendship turned to the Soviet Union being regarded in Peking as China's main enemy, a position from which there has so far been little change.

Soviet interest in South-east Asia, at first unchallenged after the Second World War when Communist insurgency was widespread, has had increasingly to compete with Chinese involvement. With large Chinese minorities in many areas, Soviet hopes have been pinned to support from other groups afraid of the Chinese. Attempts to establish Communist influence in Malaya failed through skilful British counter-measures, while Indonesia and Thailand have so far successfully combatted Communist infiltration. Growing Communist influence in the independent states established on the French withdrawal from its former Indo–Chinese territories drew the USA into a long, weary and unsuccessful campaign. Soviet support for Vietnam has been strengthened, largely to embarrass China. The Soviet Union has also established friendly relations with India, which has received considerable aid, though it has never committed itself as fully to supporting Moscow as the Soviet leadership would wish. Undoubtedly Indian fear of China has played into Soviet hands, while China has been quick to support Pakistan.

The traditional unstable political conditions and the rule in many countries of Central and South America by oppressive regimes have suggested a promising area for the spread of Soviet influence, directly within what the USA has long regarded as its own preserve. The situation looked encouraging when a successful Communist coup took over in Cuba, a considerable prestige defeat for the USA. Cuba has been set as a dubious showpiece to what might be achieved elsewhere in Latin America, while Cuban experience in insurgency was passed to many Marxist groups, though most of these movements failed. A radical left-wing state existed for a brief period in Chile until overthrown by the military, and Peru has also shown left-wing tendencies, while the left-wing position generally has been strengthened in Central America. Nevertheless, the Soviet position has been difficult because of the power of the Roman Catholic Church and because Moscow overplayed its hand in Cuba in 1962, resulting in the USA accusing the USSR of establishing a missile base on Cuba and blockading the island, so forcing Khrushchev to climb down.

In the early 1960s a considerable interest began to be taken in Africa, and friendly relations were established with some of the newly independent states in the north. Some military dictatorships with left-wing tendencies in former French territories in West Africa readily accepted aid from the USSR and Comecon members like the German Democratic Republic and Czechoslovakia. Considerable persuasion was used to generate amicable relations with the more powerful West African states like Nigeria, but an important ally was, however, found in the Congo. Much Soviet aid has been exercised through Cuban help, especially in former Portuguese territories, notably Angola. Increasingly Soviet interest in southern Africa has been faced by Chinese efforts to combat what Peking would regard as Soviet 'imperialism'. Chinese aid in building the Tanzam railway has rather outshone the material aid proffered by the Soviet Union.

Suggestions that Soviet experience rather than that of the Western industrial countries would be more relevant to solving the problems of the lesser-developed countries have been questioned. Certainly, in winning close friends among these countries, Soviet success has been modest. Perhaps this is one reason why Moscow has harnessed the only Comecon member in Latin America, Cuba, to help, since in cultural, ethnic and environmental terms it might be expected to have more feeling for Latin American and African dilemmas than the Soviet Union. It is hardly likely that Cuba could afford to pay for the aid it has given, so the presumption must be that the Soviet Union has paid the price. It could, of course, be pointed out that China, another of the less-developed countries, has afforded considerable to aid the Third World, but this might be countered by pointing to the remarkable scale of China's resources, even though still poorly harnessed, compared with those of Cuba.

The Chinese People's Republic affords an interesting example of the difficulties of applying the Soviet example to other cases. The first Chinese five-year plan (1953–7) tried to modernise and progress by following a Soviet model of the Stalinist type. Heavy industrialisation was given priority, receiving 60 per cent of investment, though 85 per cent of the population was engaged in farming. Certainly considerable achievements were made, but there was an insuperable difficulty in central direction and supervision of planning that accelerated the growth of a cumbersome bureaucracy. It was quickly obvious that China had a much greater shortage of capital resources than in the early years in the Soviet economy and that, again unlike the Soviet situation, China had vast reserves of labour that had to be absorbed, so that any development to be most effective had essentially to be as labour-intensive as possible. On the other hand, the rate at which the economy could be expanded was held down by a serious lack of skilled technicians for industry, whose numbers could not be increased until adequate training facilities had been created, a far from easy task in the Chinese setting of the period. Soviet experience in agriculture was little more applicable, because agricultural production per head in the Soviet Union in 1928 had been over twice the level in China in 1952. Even the Chinese transport system in 1952 was far inferior to what the Soviet system had been in 1928, so that the exchange of goods could not be undertaken on the scale needed to keep abreast of planned developments. Such a situation as that in China applies in many Third World countries, and Chinese experience in overcoming its own problems may consequently be more useful than the Soviet model.

9.5 The Soviet strategic position

Through the 1950s, the cold war divided the world into two major power *blocs* around the USA and the USSR as the new super-powers, but by the end of that decade both super-powers, despite their great resources, were beginning to tire of the strain of maintaining large garrison forces and competing in an accelerating technological arms race. Relations in the 1960s and 1970s were tempered by a rough working formula of 'peaceful coexistence' that produced a measure of disengagement and feelers towards *détente*, at least in Europe. Nevertheless, the talk of a bipolarity in political associations in the 1950s has retained its truth, even though the intensity of that polarity varies over the years as the pattern of tension changes.

The strategic difficulties arise from the position and configuration of Soviet territory in relation to that of its most likely opponents and competitors, the long-standing competition for political prestige with the USA and the more recent territorial and ideological confrontation with Communist China. Soviet territory stretches across the northern part of the great bicontinental mass of Eurasia, with some justification called the 'world island'. Although a strongly continental position, it has nevertheless an important relationship to the world's oceans: it should not be overlooked that a considerable proportion of the total length of the Soviet border is coastline. Until the development of modern long-range aircraft and missiles, the long frontage on the Arctic Ocean was of little military significance; now it lies across some of the most convenient great circle trajectories between Soviet Eurasia and the interior of North America, making it a highly strategic area. On the west, epicontinental seas of the Atlantic basin form the Soviet coastline, though access from them into the open ocean is dominated by potentially unfriendly powers. The eastern coastline of the Soviet Union lies behind a barrier of islands that form in a sense a 'primary coast' with the open waters of the Pacific Ocean.

The land borders of the Soviet Union lie in a broad crescentic zone of contact with highly diverse neighbouring national territories. In this 'rimland' interaction between the Socialist *bloc* and the Western or capitalist *bloc* takes place, with each searching for dominance through political and military competition. In the far north-west, a short border with Norway forms a direct contact between the Warsaw Pact countries and those of the North Atlantic Treaty Organisation. To the south, however, neutral Finland and Sweden separate the two major *blocs*, while across the European main peninsula, the continental borderland of the Soviet Union has a wide buffer zone comprising the Eastern European Comecon states. The long southern border in difficult relief lies adjacent, however, to states of uncertain allegiance, where traditionally Russian and outside interests have been in sharp competition, while the easternmost land border lies directly against Chinese territory (with the attendant problems of Sino–Soviet relations).

Long-range aircraft, missiles and satellites have greatly changed the old relationships between 'maritime' and 'continental' powers. Previously the great strength of the Russian position – lasting into the Second World War – was its vast continental distances that allowed room for free manoeuvre against an invader, while its interior lines of communication and its ability to conduct its affairs with little fear of observation were especially useful. Missiles in particular have opened the whole territory to attack, while satellites provide detailed surveillance. The fixed interior lines of communication and centres of population and industry as well as other military targets make redeployment difficult. Subject to this situation, both the Soviet Union and the USA have built up a defensive air umbrella, which on the basis of contemporary aircraft and missiles has an exceedingly broad overlap in the Arctic Ocean basin, forming a 'truce of terror' zone, and also spreads widely across both the North Atlantic and North Pacific Oceans.

While containment in the Arctic can be exercised largely through surveillance by air and satellite (with this largely empty territory surrounded by air bases and various types of early warning systems), the 'rimland' of the continental border presents more acute problems. Here, in the diverse national territories that comprise it, manoeuvre between the super-powers for prestige victories rather than clash of arms means intense diplomatic and political activity. There is nevertheless also a keen military interest, in case the brinkmanship fails and Clausewitz's dictum that war is the continuation of policy by other means becomes reality. Everywhere the pattern has been the same: the USA and its allies have tried to make the containment of the Socialist *bloc* as complete as possible, while the Soviet Union and its allies have sought to resist an encirclement that would isolate them from potential allies outside.

The rimland contact zone with the non-socialist world in Europe quickly acquired the name *Iron Curtain*, an apt description in the immediate post-war years. On the north, creation of a direct Soviet–Norwegian border in 1945 makes a particularly tender point, largely because Norways's northernmost territory covers the key strategic approaches to the big Soviet naval bases in the Kola Peninsula. The Soviet hegemony established over East Central Europe has been an uncomfortable outcome of the Second World War for the nations of Western Europe. From their base at Baltiysk-Kaliningrad, Soviet naval forces can command all the arms of the Baltic, while their garrisons in the German Democratic Republic can cover the western Baltic as well as the lower Elbe, and the Soviet garrisons command a useful forward position overlooking the Fulda Gap and the route to the middle Rhine. The territorial changes of 1945 also brought Soviet territory across the Carpathian ranges on to the Pannonian footslope, another springboard from which to dominate the Danubian and Balkan lands.

The ideological split with Yugoslavia weakened the Soviet position in South-east Europe, further worsened through the defection of Albania to the Chinese camp in 1961. No longer did the Soviet Union have a strong influence over the important Morava–Vardar corridor from the Pannonian Basin to the Aegean Sea and it lost its hold over the Straits of Otranto and the Adriatic Sea. Nevertheless, the continued loyalty of Bulgaria is a useful pressure against Turkey over the entrance to the Black Sea. Inability to win over Greece and Turkey to the Soviet side was another weakness in a key sector of the Soviet rimland, leaving a forward position for Western defence systems, themselves weakened, however, by the collapse of some 'keystone' states' loyalty to the CENTO treaty.

The most vital strategic part of the rimland lies in the Middle East, where the exceptional importance of oil resources to the capitalist economy makes it imperative for the Soviet Union to exert a strong presence. Such influence can act as a regulator to slow the pace of capitalist economic development by causing uncertainty, allowing the Socialist *bloc* an opportunity to close on the Western industrial nations' lead, while if their economic health can be impaired then their ability to keep abreast in military technology is also affected. In the Middle East the Soviet problem is to retain a happy relationship with the dominantly Islamic states, but the instability of the political scene can result in unwelcome strategic losses, as in the collapse of Soviet–Egyptian cordiality. Two particularly significant areas for the Soviet Union are Persian Azerbaydzhan (giving access to eastern Turkey and Mesopotamia) and Afghanistan – a weak state – that provides a forward position towards the entrance to the Persian Gulf and the Indus Valley, a route along which Chinese influence might also penetrate towards the Indian Ocean.

The eastern sector of the Soviet rimland, covered by the Chinese People's Republic, has become since the 1960s a peculiarly tender area, because the Chinese have expressed territorial claims on historical grounds to parts of Siberia and Soviet Central Asia. As Chinese development continues, it becomes technologically and military more formidable, posing new threats to Soviet security. Especially alarming is the excellent Chinese strategic position in missile warfare arising from the occupation of Tibet, while the strong Chinese position in Manchuria threatens the Soviet hold of the Amur-Ussuri lands (where Chinese territorial claims have some validity) and the naval base of Vladivostok, essential for the Soviet Pacific fleet. To counter this the Soviet authorities have taken a new interest in strengthening their position in Siberia, especially along the railway zone, and the reactivation of the Baykal–Amur railway project must be seen in a strategic light.

The 'Chinese problem' has given a new impetus to Soviet goodwill towards the Mongol People's Republic, a useful buffer between China and the important economic areas of Soviet Baykalia. It is also a salient position overlooking the vital Kansu corridor between the coastal and interior lands of China as well as providing a commanding position towards Peking and the Yellow Sea. The generally friendly relations between the Soviet Union and India are also coloured by considerations of its strategic value as a counter-weight to China along the Himalayan glacis. Likewise, Soviet good-will towards Vietnam and the Indo–Chinese states may also be seen in strategic terms, in outflanking China.

Since the 1950s Soviet interests have spread into the lands often described as the outer or insular crescent, notably Africa and Central and South America. To exert this influence has required development of a strategically orientated seapower, a major commitment for a state so obviously continental in its associations. One immediate difficulty arises over home bases – many historians have seen much of the territorial development of the Russian state since the time of Peter the Great as a search for warm-water ports. From Chapter 8 it will be recalled that few if any Soviet ports are not seriously troubled by winter ice. Added to this is the fact that Soviet Baltic and Black Sea ports can only be reached through narrow waters easily dominated by other powers, while the approaches to the main Far Eastern ports require movement close to Chinese and Japanese territory. Even ports open all winter under Gulf Stream influence along the Murman Coast of the Kola Peninsula, the most accessible for naval purposes all the year round, suffer from access to the Atlantic Ocean through passages between territories held by foreign powers, while weather conditions in such high latitudes in winter are exceptionally rough. Movement of Soviet naval vessels between their bases on the several Soviet seas can only be done by long circuitous voyages far from friendly bases or for a brief summer period through the narrow channel between the north Siberian coast and the Arctic pack ice.

Although Soviet naval vessels have rights to use certain friendly ports in Africa, Asia and Latin America, there are no proper naval bases on lease from other powers. Soviet naval presence therefore has to depend upon its own supply capability, but a considerable part of the fleet comprises long-range submarines that can act independently for long periods, designed for both convoy destruction and launching missiles. A big problem is that much of the approach to Soviet home naval bases can be kept under surveillance by naval forces of the NATO powers. Although the Atlantic and Pacific are covered by considerable US and NATO fleets, the Soviet navy has established a powerful presence in the Indian Ocean, around whose shores are many friendly states. It holds a powerful position at the entrance to the Red Sea, though its position is not so strong at the mouth of the Persian Gulf. From the Western point of view a strong Soviet naval hold on the Strait of Hormuz leading into the Persian Gulf or on the main channels through the island arc between the Indian Ocean and Pacific waters could be embarrassing. With Soviet naval forces nowadays active in all the oceans, the broad channel linking them together in the southern oceans is of special significance. It is not unreasonable to suggest that Soviet activity in Antarctica has more than a purely scientific nature.

9.6 Towards the year 2000

One of the most uncertain tasks is to try to predict how Soviet relations with the world at large will develop as the year 2000 approaches. Two aims of the USSR seem nevertheless assured: the maximisation of its security, and the widening of its political influence among the uncommitted nations. It is also equally certain that the avowed intent to catch up with and even overtake the Western powers in technology and economic wealth has not been abandoned. The view that capitalism will destroy itself by its own contradictions is still accepted, though in many respects there seems to be a growing congruence between the apparently opposing systems. It is unlikely that relations between the two *blocs* will become close and cordial, but there seems to be every chance that a rough but peaceful coexistence will continue and a *modus vivendi* for 'mutual advantage' established. The pointers to this are that no country in the world hoping for any reasonable technological achievement can remain out of the mainstream of international contact, exchange and trade, besides which there are also indications that the Soviet Union in its contacts with the world at large has passed the point of no return where it could once again isolate and encapsulate itself as thoroughly as it did under Stalin: perhaps even such a regime in itself would never again be possible? There are now cogent arguments to support the view that technological progress increasingly becomes less possible on a national basis alone, however great national resources may be, even for super-powers. The soaring burden of research and development costs as well as the immensity of human creativity now needed for even quite modest advances seem to have reached beyond the capability of individual economies and to have taken on an intercontinental scale. It is perhaps not far from the truth to suggest that high technology is the most potent force drawing the world together, for no nation can advance its technology sufficiently if encapsulated in itself, nor can any advance be made without such technology.

The implication of this is that the Soviet Union has passed the stage of the 'industrial revolution' and has now progressed to that of the 'technological revolution', for which its own resources in man-power, brainpower, raw materials, manufacturing capacity and investment capability are in need of outside support obtainable only in the world market-place. At the same time, it seems that the USA has also reached a point where even its own immense wealth can no longer afford the rising burden of remaining far enough ahead to be unquestioned leader in a technological race at the pace of that of the 1960s. A slackening pace does not, however, necessarily mean that the Soviet Union will catch up, particularly if it remains encapsulated just within Comecon. Technological advance has become a fully multinational research and development technique among all the most advanced Western industrial nations, whose cumulative wealth in all the relevant parameters far outmatch Comecon (let alone the Soviet Union by itself), even though the latter is a serious competitor and challenger and has already demonstrated a considerable creative potential.

The political geography of the Soviet Union by the early 1980s has become coloured by the waning of revolutionary fervour (hardly surprising when 80 per cent of the population has been born since that trauma and know no other life than the Soviet milieu) and may be why it has come to accept within Comecon some measure of the polycentrism that was such anathema in Stalin's day. On the other hand, revolutionary fervour in Communist China still ferments, resulting in exceptionally radical social and economic experiments, quite out of keeping with the Russian tradition, so that the ruthlessness of the Chinese in remodelling their society has no equivalent even in the most fervent years in the Soviet Union. It is now the Chinese side that regards Soviet ideology as an unacceptable interpretation of true Communism, resulting in the political-geographical tensions at the Sino–Soviet interface and the competition between them in the Third World. Success or failure in healing this cleft will clearly be one of the most important events in the turbulent years before the new millenium.

Despite several shocks like the Cuban crisis of 1962, the Czechoslovak crisis of 1968 and the Afghan crisis of 1980, the Soviet and Western *blocs* have settled down to a version of the 'peaceful coexistence' propagated by Khrushchev in 1958 as the ideological heat of the cold war melted away. It is unlikely that complete tranquillity on the international scene will ever be achieved, but it is equally unlikely that armed confrontation on the scale of Korea or Vietnam will again take place. If the opposing powers have managed to take the heat out of innumerable situations over the last thirty-odd years, which measured by the reactions of 1914 and 1939 could have led to global conflict, then it is unlikely they will allow themselves to drift into a position where a third world war became reality. The aftermath of the 1973 Middle Eastern crisis appeared at first to have affected only the Western industrial nations, so that the Soviet Union and Comecon, somewhat smugly, felt they were isolated from the creeping energy crisis and world-wide inflationary tendencies, but by the late 1970s it was clear they had not escaped. The growing resource crisis facing the world and the social impact of new technologies is forcing Comecon to identify its interests with those of the other advanced industrial nations. With this in mind, both *blocs* clearly recognise that *détente* and limitations on strategic armaments, at least as preliminary steps, are serious aids to releasing more of our dwindling resources for more profitable purposes.

9.7 Where to follow up this chapter

The chapter has been built upon the 'classical' view of the Russian and later Soviet geostrategic position usually accepted in the West. It is based on the concept of the world island first postulated by Sir Halford Mackinder in 'The Geographical Pivot of History', *Geographical Journal*, 23, 1904, pp.421–44, and developed further in his *Democratic Ideals and Reality* (Penguin, Harmondsworth, 1942). His views were modified and developed by Spykman, N. J., in *America's Strategy in World Politics* (Harcourt Brace, New York, 1942), and *The Geography of Peace* (Harcourt Brace, New York, 1944). The air-power aspect of Mackinder's ideas is outlined by Alexander de Seversky in *Air Power – The Key to Survival* (Simon & Shuster, New York, 1950). Also of interest is Cressey, G. B., *The Basis of Soviet Strength* (McGraw-Hill, New York, 1945). A variation is also expressed by Sir John Slessor in his *Strategy for the West* (Morrow, New York, 1954) and *The Great Deterrent* (Praeger, New York, 1957). See also Vigor, P. H., *The Soviet View of War and Peace and Neutrality* (Routledge & Kegan Paul, London, 1975). Andrei Amalrik's unusual book *Can the Soviet Union Survive Until 1984?* (revised and enlarged edition, Penguin, Harmondsworth, 1980), presents another view of the Soviet strategic position. Other variant interpretations are to be found in Meinig, D. W., 'Heartland and Rimland in Eurasian History', *Western Political Quarterly*, 9, 1955, pp. 553–69, and in Cohen, S. B., *Geography and Politics in a World Divided* (Oxford University Press, 1973), and 'The Contemporary Geopolitical Setting', in Fischer, C. A. (ed.), *Essays in Political Geography* (Methuen, London, 1968).

The involvement of the Soviet Union as the core of Comecon is discussed at length in Kaser, M., *Comecon: Integration Problems of the Planned Economies* (Royal Institute of International Affairs, Oxford, 1967), and Mellor, R. E. H., *Comecon: Challenge to the West* (Van Nostrand, New York, 1971), while the Soviet view is expressed by Meshcheryakov, V., *SEV – Printsipy, Problemy, Perspektivy* (Moscow, 1975).

Other useful sources

Bandera, V. N. and Melnyk, Z. L., *The Soviet Economy in Regional Perspective* (Praeger, New York, 1973).

Campbell, J. C., *The Soviet Union and East–West Relations* (McGraw-Hill, New York, 1979).

Dawisha, K. and Henson, P., *Soviet–East European Dilemmas* (Heinemann, London, 1981).

Dwyer, D. J., *China Today* (Longman, London, 1976).

Erikson, J., *Soviet Military Power* (Institute for Strategic Studies, London, 1971).

Freedman, L., *US Intelligence and the Soviet Strategic Threat* (Macmillan, London, 1977).

Garthoff, R. L., *Soviet Military Policy* (Faber & Faber, London, 1966).

Gorshkov, S. G., *Morskaya Mosch Gosudarstva* (Moscow, 1977).

Grechko, A. A., *Armed Forces of the Soviet Union* (Moscow, 1977).

Jackson, W. A. D., *The Russo–Chinese Borderlands* (Van Nostrand, New York, 1968).

Kinter, W. R. and Scott, H. F., *The Nuclear Revolution in Soviet Military Affairs* (University of Oklahoma Press, 1968).

Laulan, Y., *Exploitation of Siberia's Natural Resources* (NATO, Brussels, 1974).

Murphy, P. J. (ed.), *Naval Power in Soviet Policy* (Brookings Institution, Washington, D.C., 1978).

Parker, W. H., *The Super-Powers: The United States and the Soviet Union Compared* (Macmillan, London, 1972).

Possony, S. T., *Die Strategie des Friedens* (Verlag Wissenschaft und Politik, Cologne, 1964).

Sokolovsky, V. D., *Military Strategy – Soviet Doctrine and Concepts* (Faber & Faber, London, 1963).

Useful papers are also to be found in the Joint Economic Committee of the US Congress Compendium of Papers, *Soviet Economy in a Time of Change* (US Government Printing Office, Washington, D.C., 1979) 2 vols.

Bibliography

General works

Baransky, N. N., *Economic Geography of the USSR* (Progress, Moscow, 1956).

Berg, L. S., *Natural Regions of the USSR* (Macmillan, New York, 1950).

Borisov, A. A., *Climates of the USSR* (Oliver & Boyd, Edinburgh, 1965).

Cole, J. P., *A Geography of the USSR* (Penguin, Harmondsworth, 1967).

Cole, J. P. and German, F. C., *A Geography of the USSR*, 2nd edn (Butterworth, London, 1970).

Dewdney, J. C., *A Geography of the Soviet Union*, 2nd edn (Pergamon, Oxford, 1970).

Dewdney, J. C., *The USSR*, Studies in Industrial Geography (Hutchinson, London, 1978).

Gregory, J. S., *Russian Land, Soviet People* (Harrap, London, 1968).

Karger, A. and Stadelbauer, J., *Sowjetunion*, Fischer Länderkunde (Fischer Verlag, Frankfurt, 1978).

Lavrishchev, A., *Economic Geography of the USSR* (Progress, Moscow, 1969).

Mathieson, R. S., *The Soviet Union: An Economic Geography* (Heinemann, London, 1975).

Mellor, R. E. H., *Comecon: Challenge to the West*, Van Nostrand Searchlight 48 (Van Nostrand, New York, 1971).

Mellor, R. E. H., *Sowjetunion*, Harms Erdkunde (List Verlag, Munich, 1976).

Nove, A., *An Economic History of the USSR* (Penguin, Harmondsworth, 1969).

Parker, W. H., *An Historical Geography of Russia* (University of London, 1968).

Parker, W. H., *The Soviet Union*, The World's Landscapes 3 (Longman, London, 1969).

Parker, W. H., *The Super–Powers: The United States and the Soviet Union Compared* (Macmillan, London, 1972).

Pokshishevsky, V. V., *Geography of the Soviet Union* (Progress, Moscow, 1974).

Pötzsch, H. (ed.), *Die Sowjetunion*, Informationen zur Politischen Bildung 182 (Bundeszentrale für Politische Bildung, Bonn, 1979).

Symons, L. J. (ed.), *Geography of the USSR*, a series of 12 brochures (Hicks Smith, Wellington, New Zealand, 1968–70).

Useful Russian-language texts

Baranskiy, N. N., *Ekonomicheskaya Geografiya SSSR* (Moscow, 1954).
Bogush, G. M. and Shaykin, B. G., *Selskoye Khozyaystvo SSSR* (Moscow, 1977).
Cherdantsev, G. N., Nikitin, N. N. and Tutykhin, B. A., *Ekonomicheskaya Geografiya SSSR*, 3 vols (Moscow, 1956–8).
Danilov, A. D., *Ekonomicheskiye Rayony SSSR* (Moscow, 1969).
Galitskiy, M. I., *et al.*, *Ekonomicheskaya Geografiya Transporta SSSR* (Moscow, 1965).
Gladyshev, A. N., *Novyye Teritorialnyye Kompleksy SSSR* (Moscow, 1977).
Gvozdetskiy, N. A., *Fiziko-geograficheskoye Rayonirovaniye SSSR* (Moscow, 1968).
Kazanskiy, N. N. *et al.*, *Geografiya Putey Soobshcheniya SSSR* (Moscow, 1969).
Khorev, B., *Gorodskiye Poseleniya SSSR* (Moscow, 1968).
Khrushchev, A. T., *Geografiya Promyshlennosti SSSR* (Moscow, 1969).
Kopylov, I. V., *Krupnyye Ekonomicheskiye Rayony SSSR* (Moscow, 1974).
Malinin, E. D. and Ushakov, A. K., *Naseleniye Sibiri* (Moscow, 1976).
Milkov, F. N. and Gvozdetskiy, N. A., *Fizicheskya Geografiya SSSR*, 2 vols (Moscow, 1976).
Nadtochiy, G. L., *Geografiya Morskikh Putey* (Moscow, 1972).
Nikitin, N. P., Prozorov, E. D. and Tutykhin, B. A., *Ekonomicheskaya Geografiya SSSR* (Moscow, 1973).
Nikolskiy, I. V., *Geografiya Transporta SSSR*, 2nd edn (Moscow, 1980).
Pavlovskiy, E. N. (ed.), *Geografiya Naseleniya v SSSR – Osnovniye Problemy* (Moscow, 1964).
Pavlovskiy, I. G., *Problemy i Perspektivy Razvitiya Transporta* (Moscow, 1980).
Saushkin, Yu. G., *Ekonomicheskaya Geografiya Sovetskogo Soyuza*, 2 vols (Moscow, 1967–73).
Stanislavyuka, V. L., *Razvitiye Yedinoy Transportnoye Seti SSSR v Desyatoy Pyatiletke* (Moscow, 1977).
Suvorova, G. T. and Stepanov, A. I., *Ekonomicheskaya Geografiya SSSR* (Moscow, 1972).
Valentey, D. *et al.*, *Demograficheskaya Situatsiya v SSSR* (Moscow, 1976).
Zhelezko, S. N., *Sotsialno-demograficheskiye Problemy v Zone BAMa* (Moscow, 1980).

Other material

Bater, J. H., *The Soviet City*, Explorations in Urban Analysis (Arnold, London, 1979).
Billy, J. (ed.), *Comecon: Progress and Prospects* (NATO, Brussels, 1977).
Biryukov, V., 'The Baykal–Amur Mainline – A Major National Construction Project', *Soviet Geography*, 16, 1975, pp. 225–30.
Bond, A. R. and Lydolph, P. E., 'Soviet Population Change and City Growth – 1970–1979', *Soviet Geography*, 20, 1979, pp. 461–88.
Brook, S. I., 'Population of the Soviet Union – Changes in its Demographic, Social and Ethnic Structure', *Geoforum*, 9, 1972, pp. 7–21.
Clem, R. S., 'Regional Patterns of Population Change in the Soviet Union 1959–1979', *Geography Review*, 70, 1980, pp. 137–56.
Cole, U. P. and Harrison, M. E., *Regional Inequality in Services and Purchasing Power in the USSR*, Occasional Paper 14, Department of Geography, Queen Mary College, University of London, 1978.

Congress of the United States of America, *The Soviet Economy in a Time of Change*, Joint Economic Committee, 2 vols (US Government Printing Office, Washington, D.C., 1979).

Dewdney, J. C., *Patterns and Problems of Regionalisation in the USSR*, Research Papers Series 8, Department of Geography, University of Durham, 1967.

Dienes, L., 'Investment Priorities in Soviet Regions', *Annals of the Association of American Geographers*, 62, 1972.

Dienes, L., 'Basic Industries and Regional Economic Growth – the Soviet South', *Tijdschrift voor Economische en Sociale Geografie*, 68, 1977, pp. 2–15.

Dienes, L., 'Modernisation and Energy Development in the Soviet Union', *Soviet Geography*, 21, 1980, pp.121–58.

Dienes, L. and Shabad, T., *The Soviet Energy System* (Wiley, New York, 1979).

French, R. A. and Hamilton, F. E. I., *The Socialist City* (Wiley, New York, 1979).

Fuchs, R. J. and Demko, G. J., 'Commuting in the USSR', *Soviet Geography*, 19, 1978, pp.363–72.

Gerasimov, I. P. *et al.*, *Natural Resources of the Soviet Union: Their Use and Renewal* (Freeman, San Francisco, 1971).

Gibson, J. R., 'Russian Expansion in Siberia and America', *Geography Review*, 70, 1980, pp.127–36.

Gudkova, G. N. and Moskin, B. V., 'The Development of Motor Roads in the USSR', *Soviet Geography*, 15, 1974, pp.573–81.

Hamilton, F. E. I., *The Moscow City Region*, Problem Regions of Europe (Oxford University Press, 1976).

Hamilton, F. E. I., *The Planned Economies*, Aspects of Geography (Macmillan, London, 1979).

Hanson, P., 'The Soviet Energy Balance', *Nature*, 261, 1976, pp.3–5.

Harris, C. D., 'Urbanisation and Population Growth in the Soviet Union 1959–1970', *Geography Review*, 61, 1971, pp. 102–24.

Heuseler, H., *Unbekannte UdSSR aus dem All*, satellite pictures (Umschau, Munich, 1976).

Jensen, R. G., 'Soviet Regional Development Policy and the Tenth Five-Year Plan', *Soviet Geography*, 19, 1978, pp. 196- 201.

Khrushchev, A. T., 'Industrial Nodes and Principles for a Typology', *Soviet Geography*, 12, 1971, pp.91–102.

Kochetkov, A. V. and Listengurt, F. M., 'A Strategy for the Distribution of Settlement in the USSR: Aims, Problems and Solutions', *Soviet Geography*, 18, 1977, pp.660–74.

Kolosovskiy, N. N., 'The Territorial-Production Complex in Soviet Economic Geography', *Journal of Regional Science*, 3, 1961, pp.1–25.

Komar, I. V., 'The Major Economic-Geographical Regions of the USSR', *Soviet Geography*, 1, 1960, pp. 31–43.

Kovalev, S. A., 'Transformation of Rural Settlement in the Soviet Union', *Geoforum*, 9, 1972, pp.33–45.

Kurakin, A. F., 'Problems in the Spatial Concentration of Industry', *Soviet Geography*, 16, 1975, pp. 145–54.

Laulan, M. Y., *Exploitation of Siberia's Natural Resources*, NATO Round Table (NATO, Brussels, 1974).

Lis, A. G., 'On the Question of the Composition of Economic-Territorial Complexes', *Soviet Geography*, 16, 1975, pp.20–7.

Lorimer, F., *The Population of the Soviet Union: History and Prospects* (League of Nations, Geneva, 1946).

Mellor, R. E. H., 'Some Influences of Physical Environment upon Transport Problems in the Soviet Union', *Advancement of Science*, 20, 1964, pp. 564–71.

Meyer, H. E., 'The Coming Soviet Ethnic Crisis', *Fortune*, 98, 1978, pp.156–66.

Miller, E. B., 'The Trans-Siberian Landbridge: A New Trade Route between Japan and Europe – Issues and Prospects', *Soviet Geography*, 19, 1978, pp.223–43.

NATO Symposium, *Soviet Economic Growth 1970–1980* (NATO, Brussels, 1975).

Parker, W. H., *Motor Transport in the Soviet Union*, Research Paper 23, School of Geography, University of Oxford, 1979.

Rajabov, S. A., 'Geographical Factors and Certain Problems of Federalism in the USSR', *International Social Science Journal*, 30, 1978, pp.88–97.

Sallnow, J., 'The Soviet Automobile', *Geographical Magazine*, 48, 1976, pp.675–7.

Shabad, T., 'Soviet Regional Policy and CMEA Integration', *Soviet Geography*, 20, 1970, pp.230–54.

Shabad, T. and Mote, V. L., *Gateway to Siberian Resources* (Wiley, New York, 1977).

Sidorova, V. S. and Vadyukhin, A. A., 'New Technology and the Location of the Iron and Steel Industry in the Eastern Portion of the USSR', *Soviet Geography*, 18, 1977, pp.33–8.

Sochinskaya, A. V., 'Territorial-Production Complexes – An Important Approach to Cost-effective Location', *Soviet Geography*, 18, 1977, pp.374–83.

Symons, L. and White, C. (eds), *Russian Transport: An Historical and Geographical Survey* (Bell, London, 1975).

Trewartha, G. T. (ed.), *The More Developed Realm – A Geography of its Population* (Pergamon, Oxford, 1978) chs 5, 6.

Warren, K., 'Industrial Complexes in the Development of Siberia', *Geography*, 63, 1978, pp.167–78.

Yegrova, V. V., 'The Economic Effectiveness of the Construction of Pioneering Railroads in Newly Developed Areas', *Soviet Geography*, 5, 1964, pp.46–55.

Zayonchovskaya, Z. A. and Zakharina, D. M., 'Problems of Providing Siberia with Manpower', *Soviet Geography*, 13, 1972, pp.671–83.

Zimm, A., 'Zu einigen Aspekten der Standortentwicklung in der Sowjetunion im 10 Planjahrfünft', *Petermanns Mitteilungen*, 121, 1977, pp. 152–68.

Atlases

Atlas SSSR v Desyatoy Pyatilekte (Moscow, 1977).

Atlas SSSR v Devyatoy Pyatiletke (Moscow, 1972).

Atlas Razvitiya Khozyaystva i Kultury SSSR (Moscow, 1967).

Atlas SSSR (Moscow, 1969).

Maliy Ekonomicheskiy Atlas SSSR (Moscow, 1979).

USSR Agriculture Atlas (CIA, Washington, D.C., 1974).

Yearbook of statistics

Narodnoye Khozyaystvo SSSR (Central Administration for Statistics, Moscow, annual).

Index

acid production 139
Aeroflot 167
Afghanistan 29, 185, 190
age and sex of population 47ff, 70, 74
agricultural engineering 136
agricultural fertiliser 87, 89, 90, 91, 118, 120, 139
agricultural land use 85ff, 95ff
agrogorod 73, 79, 93
aircraft 98, 99, 138, 167ff
allotment gardens and plots 82, 95
alphabets in use 56
Altay mountains 4, 13, 17, 20, 64, 91, 97, 98, 118, 121
aluminium 117, 120, 133ff
Amu Darya 15, 28, 65, 92, 162
Amur 13, 32, 34, 63, 155, 157, 162, 163
Apsheron Peninsula 111, 112
Aral Sea 20, 69, 92, 112, 121, 162
Arctic *see* northlands
Arctic farms 93, 99, 100
area and extent of the USSR 1ff
Arkhangelsk 164
Armenians, Armenia 17, 29, 54, 66, 98, 116, 120
asbestos 121
Autonomous Soviet Socialist Republic 36ff
Azerbaydzhanis, Azerbaydzhan 54, 55, 65, 111, 118
Azov, Sea of 162ff
bakhchi farming 95, 99
Baku oilfield 110, 124
Baltic peoples (Latvians, Lithuanians) 55, 57
Baltic republics 29, 34, 37, 49, 53, 54, 59, 66, 69, 72, 73, 79, 95, 111, 115, 116, 124, 125, 129
Baltic (Sea and Region) 28, 29, 31, 59, 66, 72, 160ff, 162ff, 166
Baykal–Amur Railway 19, 64, 119, 157, 163
Baykalia (Cis- and Trans-) 12, 17, 21, 26, 32, 49, 55, 63, 95, 97, 99, 109, 120, 121, 129, 154, 155, 157
Baykal, Lake 4, 17, 18, 21, 63, 121, 161
Baykonur cosmodrome 64, 138
Berg, L.S., Soviet geographer 8
Bet-Pak-Dala (Hunger Steppe) 65
black earth (*chernozem*) 14, 60, 87, 95, 116
Black Sea 60, 61, 65, 67, 150, 160, 162ff
blizzards 7, 10, 12, 14, 87, 150
bogarni farming 100
Bory 11, 59, 72
Bratsk 115
Brezhnev, L., Soviet statesman 127
British capital and management in Russia 124, 125
bus services 166
Byelorussians, Byelorussia 29, 34, 39, 49, 53, 59, 67, 72, 73, 76, 95, 108, 119, 120, 140

camels 98
canals, major 160ff
Carpathian mountains 24, 31, 60, 66, 111, 121, 139
Caspian Sea 4, 6, 18, 20, 21, 66, 81, 92, 111, 120, 121, 150, 160, 162ff
cattle 93, 95, 96, 97, 98
Caucasian lands 6, 16, 17, 20, 28, 37, 39, 49, 53, 54, 55, 59, 61, 65ff, 72, 73, 76, 77, 79, 85, 91, 95, 97, 98, 99, 108, 110, 115, 120, 124ff, 128, 139, 140, 154, 155, 166, 190
CENTO Treaty 190
Central Asia 15, 20, 28, 32, 34, 37, 42, 49, 54, 64ff, 72, 76, 77, 79, 80, 85, 88, 92, 98, 109, 111, 115, 120, 124, 129, 140, 154, 156, 185ff
Central Siberian oilfields 112
Central Siberian Plateau 4, 64, 112, 166
charcoal smelting 123, 124
Chelyabinsk 131
chemicals industry 90, 107, 120ff, 122, 125, 138ff
chernozem *see* black earth
China 26, 28, 29, 32, 34, 69, 117, 156, 171ff, 178ff, 182, 184, 186, 187, 189, 190ff
Chinese and Manchurian crops in Siberia 97
Chinese Eastern Railway 155
Christianity 25, 26, 32, 54, 55, 57
chrome 117
Chukchi Peninsula 119, 120
Chuvash 53, 61
Cisbaykalia *see* Baykalia
coal, bituminous 105ff
coal, brown (lignite) 108, 109, 113
coastline 3, 12, 162
cobalt 117, 118, 119
'cold pole' of Siberia 6
Cold War 17ff, 183ff
collectivisation in farming 90ff
Comecon (Council for Mutual Economic Assistance) 31, 50, 101, 102, 116, 117, 118, 122, 127, 162, 172ff, 180ff, 183, 188
computers and related equipment 19, 44, 168, 178
concept of the 'metallurgical base' 126, 128, 131
consumer goods 123, 174ff
copper 117, 119, 133
Cossacks 26, 28, 53, 67, 68, 72, 97
cotton 92, 96, 100, 101, 141ff
creation of Soviet *imperium* in Eastern Europe 170ff
Crimea 26, 60, 69, 77, 92, 98, 99, 118, 120, 129
crop rotation 95
cross-hauls of goods 146, 149, 159
Curzon, Lord, British statesman 30

dairy farming 95ff
Dalstroy 44
Davydov Plan 20, 92, 112, 115
desalination of water 80
détente 188
distribution of national investment 123, 124, 125
Dnepr river 92, 96, 111, 115, 129, 160
Dokuchayev, .V., soil scientist 7
Donbass 61, 76, 108, 118, 125ff, 139, 155, 156
drainage and rivers 4, 6, 11, 13, 15, 16, 26, 64, 85, 90, 115ff
'drunken forest' 11
dust 13, 14, 22, 87
Dzungarian Gates 4

Eastern Siberia 49, 63, 73, 97, 109, 118, 120, 125, 129
effluent, industrial and domestic 21, 80
Ekibastuz coalfield 109, 125
electricity grids 113ff, 115, 116
electro-chemicals 139
electronics, TV and radio manufacture 136
emancipation of serfs 67, 87, 124
energy-intensive industries 105, 139, 181
engineering 61, 124, 125, 134ff
estates 88ff
Estonians 54
Eurasian 'world island' 188, 189
European Economic Community 172, 176, 183

European Russia 3, 7, 11, 13, 25, 28, 29, 31, 33, 37, 39, 49, 53, 57, 59ff, 66, 67, 72, 74, 76, 88, 95ff, 107, 108, 111, 115, 116, 118, 120, 123, 124, 125, 128, 129ff, 135ff, 139, 141, 143, 155, 159, 160, 161, 163, 165, 166, 177, 180, 182, 188, 189
exploration and colonisation 4, 12, 14, 24, 26, 28, 39, 32, 33, 53, 57, 61ff, 66ff, 79, 91

famine 90
Far East 8, 13, 19, 26, 32, 42, 49, 63, 67, 73, 76, 95, 97, 109, 118, 120, 125, 129, 162
fauna and flora 8, 10, 11, 12, 13, 15, 16, 21, 22, 26, 99
Fergana Basin 65, 92, 110, 112
Finland 28, 29, 37, 188
Finno-Ugrians 53, 54, 66, 68
firearms manufacture 125
fish 21, 100
flax 95
food-processing 100, 101
forced labour 67ff, 74, 80, 88, 90, 124
foreign investment 124, 176ff, 181
forest resources 11, 12, 13, 22, 87, 99
free market 82, 88, 90, 100
fruit cultivation 96ff

gamsil wind 16
Gazli gas fields 112
geographical pivot of history 184
Georgians, Georgia 54, 65, 108
Germans 55, 69, 124, 125
Goelro (state electrification plan) 126
gold 117, 120
Gosplan 44, 149
grain crops 95ff
greenery as amenity in towns 81, 82
gullying 15, 22, 60, 72, 87, 95, 96, 154
gypsies 55

horse-breeding 93, 97, 98
horticulture 96ff
housing 72, 73, 82, 93
hunting and breeding of fur-bearing animals 99
hydro-electricity 92, 105, 113ff, 115, 161

icebreaker *Chelyushkin* 164
incorporation of former German territory 31
Indo-Chinese border and the Soviet Union 185, 190
industry *see respective branches*
influence of world price levels on Soviet economy 181
inland waterways 80, 159ff
Interatomenergo 181
Iranians 55
iron and steel making 105, 118, 124, 125ff, 132
Iron Curtain 189
iron ore 117ff, 125ff
irrigation 14, 15, 16, 64, 65, 85, 87, 90, 92, 97, 98, 99, 115
Islam 54, 55, 56, 57, 185
izba, house type 73

Japanese aid in Siberian development 109, 119, 177, 181, 184
Jews 37, 55, 69, 124
joint development projects, with West 176ff, 181

Kalinin Commission 37, 38
Kalmyks 55, 68, 69
Kara Bogaz Gol 120, 121, 165
Karaganda industrial area 67, 80, 109, 119, 125, 128, 129, 156
Karakul (Astrakhan) sheep 93
Kara Sea 164
Karelia 29, 30, 37, 49, 53, 54, 66, 72, 115, 118, 120, 121, 129
Kazakhs, Kazakhstan 37, 54, 64, 67, 69, 73, 76, 91, 109, 112, 119, 120, 125, 129, 138, 139, 140
Khrushchev, N., Soviet statesman 43, 73, 90, 127, 138
Kiev 24, 25, 26, 60, 61, 125
Kirgiz, Kirgizia 54, 120
Komsomolsk 127
kray 37ff
kreml (kremlin) 79, 81
Krivoy Rog 118, 124, 130
Kuban district 61, 72, 92, 97, 111

Kursk Magnetic Anomaly 60, 118, 129, 131, 177
Kuzbass 18, 63, 109, 125ff, 128, 156

labour supply 50, 57, 62, 63, 68ff, 74, 84, 87, 88, 90, 123, 124, 125, 157, 171, 181ff
landlord 88, 89, 90
land reclamation: in the forest zone 91, 95; in the steppe 91ff, 95, 97
lead and zinc 117, 119, 120, 133
Lena river 10, 12, 95, 97, 112, 157, 161, 164
Lenin, V.I., Soviet statesman 39, 113, 157
Leningrad 30, 59, 76, 95, 116, 124, 125, 129, 139
lignite *see* coal (brown)
livestock 89, 90, 91, 95ff
loess 14, 15, 16, 65, 88
Lysenko, T., Soviet geneticist 20, 91

Machine and Tractor Stations 92
machine-tool manufacture 136
Mackinder, Sir H. J., British geographer 3, 184
magnesium 117, 120
Magnitogorsk 118, 126, 131
maize 95ff
makrorayon 39
Manchuria 28, 29, 32, 63, 69, 155, 156, 185
manganese 117, 118, 119, 130
Mangyshlak Peninsula 64, 80, 112, 156, 164
Manych Depression 92
Mari 54, 96
'maritime' and 'continental' power in a Soviet setting 189, 191, 192
Marshall Plan aid 172
Marxism-Leninism 20, 32, 73, 74, 79, 81, 83, 90, 92, 113, 122ff, 126, 143, 172ff
mercury 117, 119
Meshchera Depression 59, 96
mica 121
millet 96, 97
mining equipment and steel-making machinery manufacture 135ff
mixed farming 95ff
Moldavians, Moldavia 37, 49, 55, 60, 96
molybdenum 117
Mongols, Mongolia 25, 26, 32, 53, 54, 55, 63, 99, 185, 191
Mordovs 54, 61
Moscow, city and region 26, 28, 30, 32, 59, 76, 79, 80, 95, 105, 108, 115, 124, 129, 135, 136ff, 138, 140, 155, 156
Murmansk 164

Nakhodka 165
nationalities 33ff, 36, 37, 53ff
'national principle' 39, 53, 56, 57, 70
NATO (North Atlantic Treaty Organisation) 188, 192
natural gas 105ff, 115, 139, 140
navigation on Siberian rivers 161
Nebit Dag oilfields 112, 139
New Economic Policy 39, 90, 126
'New Russia' 14, 26, 60, 61, 67, 72
nickel 117, 119
Nizhniy Tagil 131
nomads 24, 25, 53, 54, 55, 59, 64, 65ff, 71, 72, 98
Northern Caucasia *see* Caucasian lands
Northern Caucasian oilfields 111
Northern Sea Route 30, 163, 164ff, 191ff
northlands and the Arctic 64, 68, 72, 76, 80, 93, 99, 111, 120, 157, 172, 188
Novgorod 25, 26, 79
Novosibirsk 63, 80, 116, 139
nuclear energy 80, 105, 113ff, 116, 164, 181

oases 95, 99
oblast 36ff
offshore oil resources 110
oil-refining 111, 139, 140
oil resources 105ff, 109, 110, 115, 139
oil seeds 97, 99
oil shale 105, 113, 115
Okhotsk Sea 163
optimal size of towns 82

Orthodox Church *see* Christianity
Outer Mongolia, Mongol People's Republic *see* Mongols, Mongolia

Paleao-Asiatics 55
Pan-Slavism 56
pastoralism 85, 91, 95, 97ff
peat 105, 115
Pechenegs and early nomadic invaders 25
Pechora coalfield 108
permafrost 6, 8, 10, 11, 19, 80, 87, 93, 112, 150, 155
Persian Azerbaydzhan 185, 190
Peter I, the Great 28, 123, 191
petrochemicals 138ff
pigs 95, 96, 98
pipelines, oil and gas 110ff
platinum 117
podzol 10, 13, 87, 93, 95
Polesye, Pripyat marshes 13, 24, 59, 66, 72, 91
Poland 26, 28, 29, 30, 34, 53
polymetallic ores 119, 133ff
port facilities 163ff, 191
portages 79
potatoes 95
power-generating equipment manufacture 136
precipitation 7, 13, 14, 16, 85, 87, 91
public-health problems 13, 21
purchase of technology, etc., from the West 174ff

railway engineering 124, 125, 136ff
railway gradients 154
railway motive power problems 154ff, 157ff
railways, dates of opening of major lines 155ff
rasputitsa 7, 14
regional planning 107, 113, 122ff, 126ff, 138, 143ff, 149, 180
regional variations in farming 93ff
reindeer 99
relief 4, 5, 6, 16, 152ff
reparations from Germany after 1945 127
rice 96ff
road–rail transport relations 165
roads, length and type 166
road transport 81, 165ff
rubber 100, 125, 140
Russian Federation (Russian Soviet Federated Socialist Republic) 37, 39, 49, 56
Russian Poland 124
Russian spread into North America 29
Rustavi 118
rye 95, 97

Sakhalin 109, 111, 112, 157, 163
salts 120ff, 125, 139, 140
Sayan mountains 17, 64, 99
'scorched-earth' policy 171, 143
search for warm-water ports 29, 162, 191
seasonal fairs in Old Russia 79, 124, 125
seasons, influence on human activity 6, 14, 85ff
'Second Baku' oilfields 110ff, 139
self-sufficiency 102, 105, 122ff, 127, 140, 143, 149, 173, 180ff
Semirechye 15
service frequency on Soviet railways 158ff
settlements of town type (workers' settlements, etc.) 74, 77
sheep 93, 97, 98
shipbuilding 137
Siberia 3, 4, 6, 8, 11, 12, 13, 17, 18, 19, 20, 26, 28, 31, 49, 53, 55, 57, 59, 62ff, 66ff, 71, 74, 76, 80, 85, 87, 89, 91, 95, 97, 98, 99, 105, 109, 111, 112, 113, 115, 116, 118, 120, 124, 125ff, 140, 155ff, 166, 180ff
Siberia, conquest of 26, 28, 34, 53, 55, 57, 67ff
Sibiryak 53, 57
silk 100, 141
silver 117, 118, 120, 133
Slavs 24, 25, 26, 28, 32, 33, 34, 53ff, 66ff, 72
Smolensk gateway 26, 30, 59
snow 7, 10, 13, 87, 97
soils 8, 10, 11, 12, 13, 14, 15, 16, 17, 65, 87ff, 94ff
Soviet grain purchases in the West 176

Soviet interest: in Africa 178ff, 191; in Egypt 184ff, 190; in Latin America 186, 187, 191; in Middle East 190, 191; in South-east Asia 185ff, 191; in world sea routes 182ff, 188ff, 192
Soviet interior lines of communication 189
Soviet of Nationalities, of the Union 36ff
Soviet Pacific coast 26, 32, 138, 163ff, 188, 191
Soviet relations: with the Third World 176ff, 191ff; with the USA 171, 176, 177, 178, 182, 186, 188ff, 192; with Yugoslavia 190
Soviet strategic position in Central Europe 189, 190
Soviet trading system 173, 175ff, 176
Soviet world position 3, 4, 18, 19, 29, 181ff, 191
Sovietisation 31, 32, 34, 57, 79, 175
sovnarkhoz 43
space research and missile technology 172, 178, 189
Stalin, J.V., Soviet statesman 31, 81, 90, 91, 122, 127, 157, 170, 171ff
state farms 93
Stavropol natural-gas field 111
Stolypin reforms 89
Strait of Kerch 163
strategic role of the Arctic 172, 191, 192
Stroganov family 124
suburban farming 95, 100
sugar beet 96ff, 101
sukhovey wind 14, 87, 91
sulphur 120, 139
sunflower cultivation 95ff
Sverdlovsk 131
Swedish territories taken in the Baltic 28
synthetic yarns manufacture 139, 141
Syr Darya 28, 65, 162

Tadzhiks, Tadzhikstan 55, 110, 112, 140
Tashkent 65, 110, 136
Tatars 53, 61
Taymyr Peninsula 109, 111, 112
tea 99
technology gap with the West 171ff, 181, 192ff
territorial production complex 44, 123, 143ff, 149
textile manufacture 59, 124, 125, 135, 141ff
thermal electricity 108, 113ff
thermo-plastics manufacture 140
time zones 19, 116, 168
tin 117, 120
tobacco and makhorka 96
town as a 'proletarian base' 73
'town of socialist realism' 81
Trans-Aral Railway 155
Transbaykalia *see* Baykalia
Trans-Caspian Railway 155
Transcaucasia *see* Caucasian lands
'transformation of nature' 20, 92, 98, 112
transport system, dimensions of 150
Trans-Siberian Railway 62ff, 67, 68, 77, 97, 155ff, 183
tugai 15
tungsten 117, 119
Tungus-Manchurians 55
Turkey 28, 29, 190
Turkic peoples 53ff
Turkmen, Turkmenistan 54, 65, 92, 112, 120
Turksib Railway 156
Tuva 32, 98
Tyan Shan mountains 4, 16
types of land tenure in Old Russia (*barshchina*, *obrok*) 88

Udmurts 54
Ukhta oilfield 111, 139
Ukrainians, Ukraine 29, 34, 37, 39, 42, 49, 53, 59, 60ff, 67, 72, 76, 80, 92, 96, 108, 111, 118, 120, 121, 124, 125, 128, 129, 135, 139, 155
unified transport system 146ff
Union Republics 36ff, 53ff
unrealised railway projects 157
Ural, region and mountains 4, 11, 17, 18, 22, 26, 39, 42, 49, 62, 66, 67, 76, 80, 97, 105, 107,

108, 109, 110, 118, 119, 120, 121, 123, 124, 125, 128, 129ff, 135, 152, 155
Ural–Emba oilfield 112
Ural–Kuzbass *Kombinat* 126ff, 128ff, 156
Ural–Volga oilfields *see* 'Second Baku' oilfields
urban 'dead heart' 81
urban hierarchy 77
use of low-grade fuels 115
Ust-Urt Plateau 64
Uzbeks, Uzbekistan 54, 65, 110, 112

Valday Hills 60, 66
vanadium and titanium 117, 118, 119
Varangians 25
village commune (*mir*) 88, 89
village industry 124
village types 72, 92, 93
vines 96ff
Virgin Lands 20, 22, 49, 64, 67, 73, 79, 87, 91, 93, 95ff
Vladimir-Suzdal 25
Vladivostok 125, 155, 191
Volga, region and river 20, 22, 26, 53, 54, 61, 66ff, 72, 76, 77, 80, 92, 95, 97, 110, 115, 120, 126, 129, 139, 155, 159ff, 185

wartime shifts in industry 127, 143
Warsaw Pact 188
water resources 21, 80, 115ff
weather forecasting, role of 98, 164
Western Siberia 11, 20, 49, 62, 63, 67, 80, 97, 109, 110, 111, 112, 115, 116, 118, 125, 157, 166
Western Siberian oilfields 112, 115, 156, 181
wheat 95ff
wooded steppe 14, 25, 66, 87, 91, 95
wooded tundra 10

yaks 99
Yakuts 68, 97
Yenisey river 10, 32, 80, 93, 115, 157, 161, 164
yurt (felt tent) 72

Zhdanov 118, 130
zinc *see* lead and zinc